Calculus

IBHL Topic 5

Analysis and Approaches

Steven Clarke
United Nations International School
New York

This book covers all of the material required for Topic 5 (Calculus) in the new syllabus of IB Higher Level Mathematics (Analysis and Approaches).

The book can be used for self-study, as numerous worked examples are provided.

If you find any typographical errors, or incorrect answers, please let me know at unismathSC@gmail.com

FIRST EDITION

ISBN : 979-8841079163

Contents

Differentiation

Integration

Differential equations

Maclaurin

Answers

Also available in the same series :

ISBN : 979-8429625102

Differentiation

$\frac{dy}{dx}$ is the most common symbol of differentiation.

It indicates "the rate of change of y compared to x".

In simple terms this is $\frac{\text{change in } y}{\text{change in } x}$

If x and y form coordinate axes then this represents gradient or slope.

If s = distance in meters and t = time in seconds

then $\frac{ds}{dt}$ indicates " the rate of change of *distance* compared to *time* ".

This is speed or velocity.

If V = volume and t = time then $\frac{dV}{dt}$ indicates "the rate of change of *volume* compared to *time*".

For example, when blowing up a balloon, if the volume is increasing at 5 cm^3 per second then you can write $\frac{dV}{dt} = 5$ cm^3/sec

First differentiation formula :

If $y = ax^n$ where $a \in \mathbb{R}$, $n \in \mathbb{R}$ then

$$\frac{dy}{dx} = nax^{n-1}$$

Examples : If $y = 4x^3$ then $\frac{dy}{dx} = 12x^2$

If $y = 10x^5$ then $\frac{dy}{dx} = 50x^4$

If $y = 4x^{\frac{1}{2}}$ then $\frac{dy}{dx} = 2x^{-\frac{1}{2}}$

If $y = 9x^{-\frac{1}{3}}$ then $\frac{dy}{dx} = -3x^{-\frac{4}{3}}$

If $y = 5x$ then $\frac{dy}{dx} = 5$

(this is because if $y = 5x^1$ then $\frac{dy}{dx} = 5x^0$)

If $y = 8$ then $\frac{dy}{dx} = 0$

(this is because $y = 8$ is a horizontal line which has zero gradient)

There are other ways to indicate differentiation :

$\frac{d}{dx}\left(4x^3\right) = 12x^2$

If $f(x) = 4x^3$ then $f'(x) = 12x^2$

Example :

Given $f(x) = \dfrac{18}{\sqrt[3]{x}}$ calculate $f'(x)$ and hence find $f'(27)$

Step 1 : Rewrite the function in the form ax^n

$$f(x) = \frac{18}{\sqrt[3]{x}}$$

$$= 18x^{-\frac{1}{3}}$$

Step 2 : Differentiate

$$f'(x) = -6x^{-\frac{4}{3}}$$

$$= \frac{-6}{x^{\frac{4}{3}}}$$

$$= \frac{-6}{\sqrt[3]{x^4}}$$

Step 3 : Substitute $x = 27$

$$f'(x) = \frac{-6}{\sqrt[3]{x^4}}$$

$$f'(27) = \frac{-6}{\sqrt[3]{27^4}}$$

$$= -\frac{6}{81}$$

$$= -\frac{2}{27}$$

Consider the straight line $y = 2x + 4$

Here $\dfrac{dy}{dx} = 2$ and this is the slope of the line.

Note : $y = 4 \Rightarrow \dfrac{dy}{dx} = 0$

The line $y = 4$ is parallel to the x-axis and hence horizontal. It has zero slope.

Gradient of a curve

A straight line has a fixed gradient.
The gradient of a curve is always changing – it depends upon which point on the curve you are considering.

The gradient of a curve at any point on the curve is defined as the gradient of the tangent to the curve at that particular point.

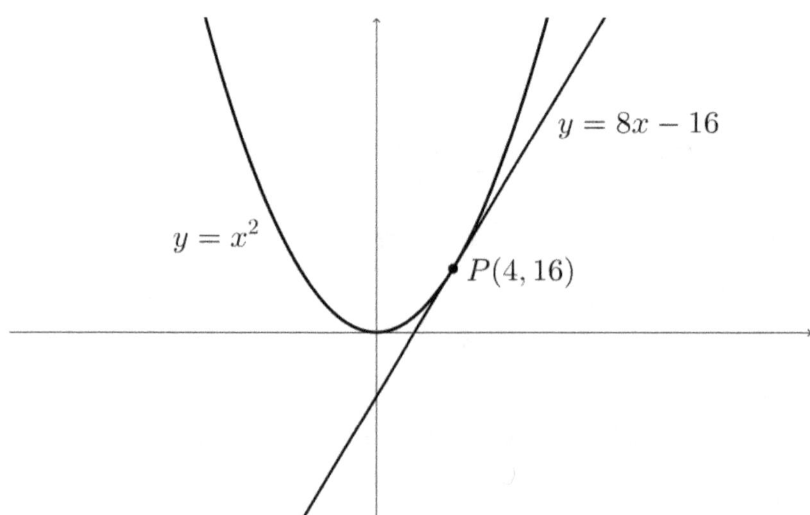

The gradient of the curve $y = x^2$ at the point P is 8
This is the gradient of the tangent shown on the graph.

Example :

Find the equation of the tangent to the curve $y = x^2$ at the point $(-1\ ,1)$

Step 1 : Find $\dfrac{dy}{dx}$ (the gradient function of the curve)

$$y = x^2 \Rightarrow \frac{dy}{dx} = 2x$$

Step 2 : Substitute $x = -1$ (to find the gradient of the curve at the required point and hence the gradient of the tangent)

When $x = -1$ $\dfrac{dy}{dx} = -2$

Step 3 : Use $y = mx + c$ to find the equation of the tangent

$$y = mx + c \Rightarrow y = -2x + c$$

When $x = -1$ $y = 1$

Hence $1 = -2(-1) + c$

$c = -1$

Therefore the equation of the tangent is $y = -2x - 1$

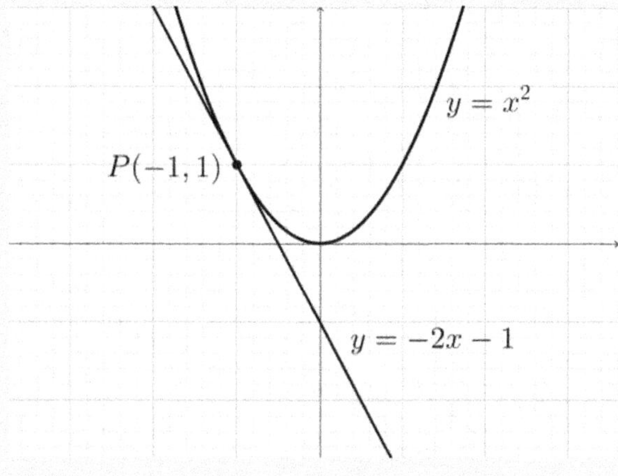

Example :

Find the equation of the normal to the curve $y=\frac{1}{8}x^3$ at the point $(2\,,1)$

Give your answer in the form $ax+by+d=0$ where $a,b,d \in \mathbb{Z}$

Step 1 : Find $\frac{dy}{dx}$ (the gradient function of the curve)

$$y=\frac{1}{8}x^3 \;\Rightarrow\; \frac{dy}{dx}=\frac{3}{8}x^2$$

Step 2 : Substitute $x=2$ (to find the gradient of the curve at the required point and hence the gradient of the tangent)

When $x=2$, $\frac{dy}{dx}=\frac{3}{2}$

If the gradient of the tangent is $\frac{3}{2}$

then the gradient of the normal is $-\frac{2}{3}$

Step 3 : Use $y=mx+c$ to find the equation of the normal

$$y=mx+c \;\Rightarrow\; y=-\frac{2}{3}x+c$$

When $x=2$, $y=1$

Hence $1=-\frac{2}{3}(2)+c$

$$c=\frac{7}{3}$$

Therefore the equation of the tangent is

$$y=-\frac{2}{3}x+\frac{7}{3}$$

$$3y=-2x+7$$

$$2x+3y-7=0$$

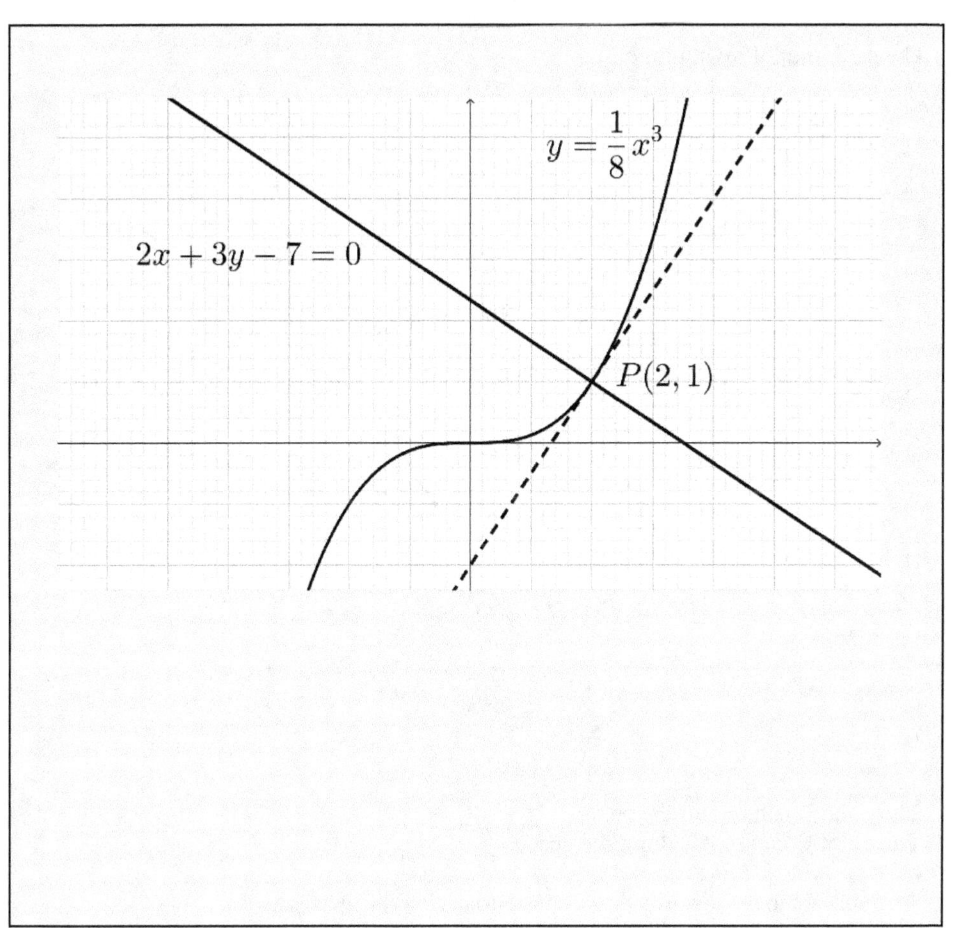
$y = \frac{1}{8}x^3$
$2x + 3y - 7 = 0$
$P(2, 1)$

Consider the following graph :

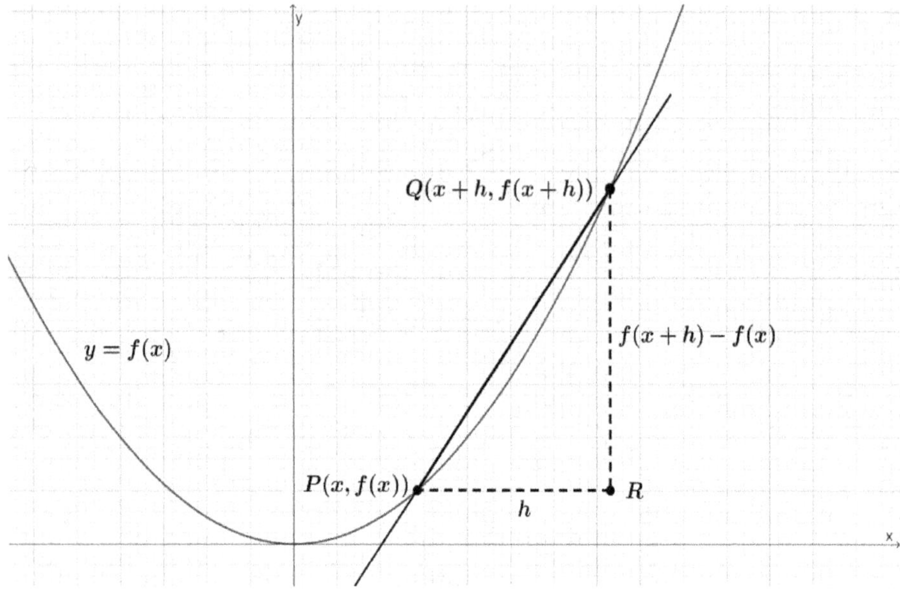

The point P is fixed on the curve.
The point Q can move along the curve.

As point Q moves towards point P , the triangle PQR gets smaller.

The gradient of PQ is given by $\dfrac{\text{change in } y}{\text{change in } x}$ or $\dfrac{f(x+h)-f(x)}{h}$

The closer Q approaches P , the closer h approaches zero , and the closer the gradient of PQ approaches the gradient of the tangent at point P.

Hence the gradient of the tangent at point P is the *limit* of

$$\frac{f(x+h)-f(x)}{h} \quad \text{as} \quad h \to 0$$

This is written as $\lim\limits_{h\to 0}\left[\dfrac{f(x+h)-f(x)}{h}\right]$

Differentiation from First Principles :

$$\text{If } y = f(x) \text{ then } \frac{dy}{dx} = \lim_{h \to 0}\left[\frac{f(x+h) - f(x)}{h}\right]$$

Example : Given $y = x^2$ show that $\frac{dy}{dx} = 2x$

Let $y = f(x)$

$$\frac{dy}{dx} = \lim_{h \to 0}\left[\frac{f(x+h) - f(x)}{h}\right]$$

$$= \lim_{h \to 0}\left[\frac{(x+h)^2 - x^2}{h}\right]$$

$$= \lim_{h \to 0}\left[\frac{x^2 + 2xh + h^2 - x^2}{h}\right]$$

$$= \lim_{h \to 0}\left[\frac{2xh + h^2}{h}\right]$$

$$= \lim_{h \to 0}\left[2x + h\right]$$

$$= 2x$$

Example : Given $y = ax^n$ show that $\dfrac{dy}{dx} = nax^{n-1}$

Let $y = f(x)$

$$\frac{dy}{dx} = \lim_{h\to 0}\left[\frac{f(x+h)-f(x)}{h}\right]$$

$$= \lim_{h\to 0}\left[\frac{a(x+h)^n - ax^n}{h}\right]$$

$$= \lim_{h\to 0}\left[\frac{a\left(x^n + nx^{n-1}h + \frac{n(n-1)}{2!}x^{n-2}h^2 + \dots\dots\dots + h^n\right) - ax^n}{h}\right]$$

$$= \lim_{h\to 0}\left[\frac{ax^n + anx^{n-1}h + \frac{an(n-1)}{2!}x^{n-2}h^2 + \dots\dots\dots + ah^n - ax^n}{h}\right]$$

$$= \lim_{h\to 0}\left[\frac{anx^{n-1}h + \frac{an(n-1)}{2!}x^{n-2}h^2 + \dots\dots\dots + ah^n}{h}\right]$$

$$= \lim_{h\to 0}\left[anx^{n-1} + \frac{an(n-1)}{2!}x^{n-2}h + \dots\dots\dots + ah^{n-1}\right]$$

$$= anx^{n-1}$$

$$= nax^{n-1}$$

The Chain Rule

$$\frac{dy}{dx} = \frac{dy}{du} \bullet \frac{du}{dx}$$

For example , if $y = 4u^2$ and $u = x^3$ then

$$\frac{dy}{dx} = \frac{dy}{du} \bullet \frac{du}{dx}$$

$$= 8u \bullet 3x^2$$

$$= 24ux^2$$

$$= 24x^5$$

Here this can easily be confirmed by substituting $u = x^3$ into $y = 4u^2$ to give $y = 4x^6$ and hence $\frac{dy}{dx} = 24x^5$.

Since $\frac{dy}{dx} \bullet \frac{dx}{dy} = 1$ we have $\frac{dy}{dx} = \frac{1}{\frac{dx}{dy}}$

Example : Given $y = 3t^5$ and $x = 8t^2$ find $\frac{dy}{dx}$ when $t = 2$

Step 1 : Find $\frac{dy}{dt}$ and $\frac{dx}{dt}$

$$y = 3t^5 \Rightarrow \frac{dy}{dt} = 15t^4$$

$$x = 8t^2 \Rightarrow \frac{dx}{dt} = 16t$$

Step 2 : Find $\frac{dt}{dx}$

$$\frac{dx}{dt} = 16t \Rightarrow \frac{dt}{dx} = \frac{1}{16t}$$

Step 3 : Find $\frac{dy}{dx}$

$$\frac{dy}{dx} = \frac{dy}{dt} \bullet \frac{dt}{dx}$$

$$= 15t^4 \bullet \frac{1}{16t}$$

$$= \frac{15t^3}{16}$$

$$= \frac{15}{2} \text{ when } t = 2$$

Example : The volume of a balloon is increasing at 72 cm^3 per second.

(a) Find the rate of change in the radius of the balloon when the volume is 36π cm^3.

(b) Find the corresponding rate of change in the surface area of the balloon (when the volume is 36π cm^3).

Step 1 : $\frac{dV}{dt} = 72$ cm^3/sec

Step 2 : Find $\frac{dV}{dr}$ (the rate of change of the volume compared to the radius)

For a sphere $V = \frac{4}{3}\pi r^3$

Hence $\frac{dV}{dr} = 4\pi r^2$

Step 3 : Find $\frac{dr}{dt}$

$$\frac{dr}{dt} = \frac{dr}{dV} \bullet \frac{dV}{dt}$$

$$= \frac{1}{4\pi r^2} \bullet 72$$

$$= \frac{18}{\pi r^2}$$

Step 4 : Find the radius when the volume is $36\pi\ \text{cm}^3$

$$V = \frac{4}{3}\pi r^3$$

$$36\pi = \frac{4}{3}\pi r^3$$

$$r = 3\ \text{cm}$$

Step 5 : Substitute $r = 3$ cm into $\frac{dr}{dt}$

$$\frac{dr}{dt} = \frac{18}{\pi r^2}$$

$$= \frac{2}{\pi}\ \text{cm per sec}$$

Step 6 : Find $\frac{dA}{dr}$ (the rate of change of the surface area compared to the radius)

For a sphere $A = 4\pi r^2$

Hence $\frac{dA}{dr} = 8\pi r$

Step 7 : Find $\frac{dA}{dt}$ and substitute $r = 3$ cm

$$\frac{dA}{dt} = \frac{dA}{dr} \bullet \frac{dr}{dt}$$

$$= 8\pi r \bullet \frac{2}{\pi}$$

$$= 16r$$

$$= 48\ \text{cm}^2\ \text{per sec}$$

Exercise 1

1) Find $\frac{dy}{dx}$

(a) $y = 9x^5$ (b) $y = x^3 + x^2 + x + 5$ (c) $y = 8\sqrt{x}$

(d) $y = \frac{2}{x^3}$ (e) $y = \frac{6}{\sqrt{x}}$ (f) $y = (2x+1)^2$

2) Find $f'(x)$

(a) $f(x) = \frac{1}{x}$ (b) $f(x) = 15\sqrt[3]{x^2}$ (c) $f(x) = \pi x$

(d) $f(x) = \frac{3x^2 + 4x - 2}{x}$

3) (a) Find the gradient of the curve $y = 2x^3 + 3x^2$ when $x = 1$

(b) Hence find the equation of the tangent to the curve when $x = 1$

4) (a) Find the gradient of the curve $y = \frac{1}{x^2}$ when $x = 2$

(b) Hence find the equation of the normal to the curve when $x = 2$

5) An object moves such that $s = 4t^3 - 3t^2$ where s = distance from the origin and t = time in seconds

(a) Find v where $v = \frac{ds}{dt}$

(b) Hence find the speed of the object after 2 seconds.

(c) Find the times when the speed is zero.

(d) Find $\frac{dv}{dt}$ and describe what you are finding.

6) Given $y = 4\sqrt{t}$ and $x = \dfrac{4}{3t^2}$, find $\dfrac{dy}{dx}$ when $t = 4$

7) The radius of a sphere is increasing at 2 cm per second. Find the rate of increase in the volume of the sphere when the diameter is 10 cm.

8) Given that $y = 2x + 1$ show from first principles that $\dfrac{dy}{dx} = 2$

9) Given that $y = x^3$ show from first principles that $\dfrac{dy}{dx} = 3x^2$

10) Use the chain rule to show that $\dfrac{dv}{dt}$ can be written as $v\dfrac{dv}{ds}$

where $v =$ velocity, $s =$ distance, $t =$ time.

11) The volume of a cube is increasing at 6 cm^3 per second. Find the rate of increase in the length of the side of the cube when the volume is 125 cm^3.

12) A hollow cone has vertical height 20 cm and base radius 4 cm. The cone is inverted and water is poured in at a rate of 2 cm^3 per second. Find the rate of change in the depth of the water when the depth is 10 cm.

13) Grain is continuously poured out of a hopper at a rate of 200 cm^3 per second onto level ground. The grain forms a cone that has the same vertical height as its base diameter. Calculate the rate of change in the height of the cone when the base radius is 50 cm.

Stationary Points

Consider the following graph of a cubic function :

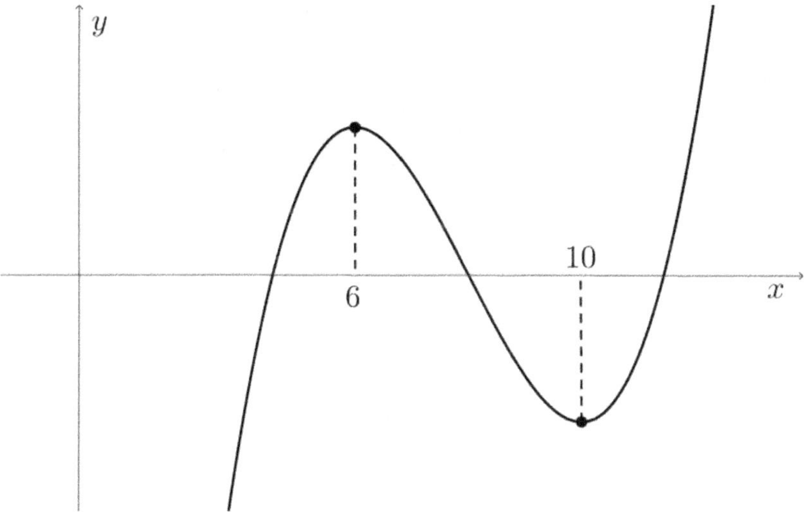

When $x < 6$ the slope of the graph is positive and hence $\frac{dy}{dx} > 0$
The function is said to be ***increasing*** .

When $6 < x < 10$ the slope of the graph is negative and hence $\frac{dy}{dx} < 0$
The function is said to be ***decreasing*** .

When $x > 10$ the slope of the graph is again positive and the function is again *increasing* .

When $x = 6$ the slope of the graph is zero and hence $\frac{dy}{dx} = 0$
This is called a *stationary point*. Here it is also a *local maximum point.*

When $x = 10$ the slope of the graph is again zero and hence $\frac{dy}{dx} = 0$
This is also *stationary point* , however now it is a *local minimum point.*

Example : Given the function $y = x^3 - 9x^2 + 24x + 5$ calculate the coordinates of the stationary points and prove their nature (i.e. local maximum or local minimum) algebraically

Step 1 : Find $\frac{dy}{dx}$ (the gradient function of the curve)

$$y = x^3 - 9x^2 + 24x + 5$$

$$\frac{dy}{dx} = 3x^2 - 18x + 24$$

Step 2 : Solve $\frac{dy}{dx} = 0$

$$3x^2 - 18x + 24 = 0$$
$$x^2 - 6x + 8 = 0$$
$$(x-2)(x-4) = 0$$
$$x = 2, 4$$

Step 3 : Calculate the y-coordinates

When $x = 2$ $y = 25$
When $x = 4$ $y = 21$

Step 4 : Construct a table showing the slope just before and just after a stationary point

x	1	2	3
$\frac{dy}{dx}$	9	0	-3

Here the slope is changing from positive to negative. This indicates a local maximum point.

x	3	4	5
$\frac{dy}{dx}$	-3	0	9

Here the slope is changing from negative to positive.
This indicates a local minimum point.

Step 4 : Write down the answers :

(2 , 25) local maximum

(4 , 21) local minimum

Higher differentials

If $f(x) = y$ then $f'(x) = \frac{dy}{dx}$

$$f''(x) = \frac{d}{dx}\left(\frac{dy}{dx}\right) = \frac{d^2y}{dx^2}$$

$$f'''(x) = \frac{d}{dx}\left(\frac{d^2y}{dx^2}\right) = \frac{d^3y}{dx^3}$$

Hence if $y = 5x^4$ then $\frac{dy}{dx} = 20x^3$

$$\frac{d^2y}{dx^2} = 60x^2$$

$$\frac{d^3y}{dx^3} = 120x$$

Points of Inflection and Concavity

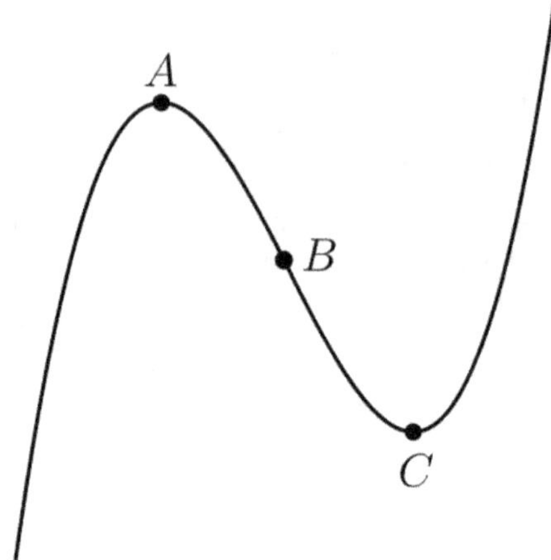

The point B is called a *point of inflection.*
We say there is a *change in concavity* at point B.
The concavity changes from negative to positive.

The curve to the left of point B is *concave down.*

The curve to the right of point B is *concave up.*

At point B we have $\dfrac{d^2y}{dx^2} = 0$

At point A we have $\dfrac{dy}{dx} = 0$ and $\dfrac{d^2y}{dx^2} < 0$

At point C we have $\dfrac{dy}{dx} = 0$ and $\dfrac{d^2y}{dx^2} > 0$

Concave down $\Rightarrow \dfrac{d^2y}{dx^2} < 0$

Concave up $\Rightarrow \dfrac{d^2y}{dx^2} > 0$

Example : Given the function $y = x^3 - 9x^2 + 24x + 5$ calculate the coordinates of the point of inflexion. Justify your answer.

Step 1 : Find $\dfrac{d^2y}{dx^2}$

$$y = x^3 - 9x^2 + 24x + 5 \Rightarrow \frac{dy}{dx} = 3x^2 - 18x + 24$$

$$\frac{d^2y}{dx^2} = 6x - 18$$

Step 2 : Solve $\dfrac{d^2y}{dx^2} = 0$

$$6x - 18 = 0 \Rightarrow x = 3$$

Step 3 : Calculate the y-coordinate

When $x = 3 \quad y = 23$

Hence the point of inflexion is (3, 23)

Step 4 : Construct a table showing $\dfrac{d^2y}{dx^2}$ just before and just after the point of inflexion

x	2	3	4
$\dfrac{d^2y}{dx^2}$	− 6	0	6

$\dfrac{d^2y}{dx^2}$ is changing from negative to positive.

This indicates a change in concavity.
Hence $x = 3$ is a point of inflexion

Concavity can sometimes be used as a short-cut to determine the nature of stationary points :

If $\dfrac{d^2y}{dx^2} > 0$ at a stationary point then the curve is concave up and hence the stationary point is a local minimum.

If $\dfrac{d^2y}{dx^2} < 0$ at a stationary point then the curve is concave down and hence the stationary point is a local maximum.

If $\dfrac{d^2y}{dx^2} = 0$ at a stationary point then you must revert back to the method of constructing a table showing the slope just before and just after the stationary point.

There are **two** types of point of inflexion :

Stationary and **Non-stationary**

As seen before , the curve $y = x^3 - 9x^2 + 24x + 5$ has a non-stationary point of inflexion at the point (3, 23)

Consider the curve $y = x^3$

$\dfrac{dy}{dx} = 3x^2$ which implies a stationary point at the origin (0,0)

$\dfrac{d^2y}{dx^2} = 6x$ which means that when $x = 0$, $\dfrac{d^2y}{dx^2} = 0$

Hence we need to revert back to the method of constructing a table to determine the nature of the stationary point.

x	-1	0	1
$\frac{dy}{dx}$	3	0	3

Since the slope is positive both before and after the stationary point ,
it must be a point of inflexion.

This can be confirmed by constructing a table showing $\frac{d^2y}{dx^2}$ just before and just after the point.

x	-1	0	1
$\frac{d^2y}{dx^2}$	-6	0	6

$\frac{d^2y}{dx^2}$ is changing from negative to positive. This indicates a change in concavity.

Hence $x = 0$ is a point of inflexion

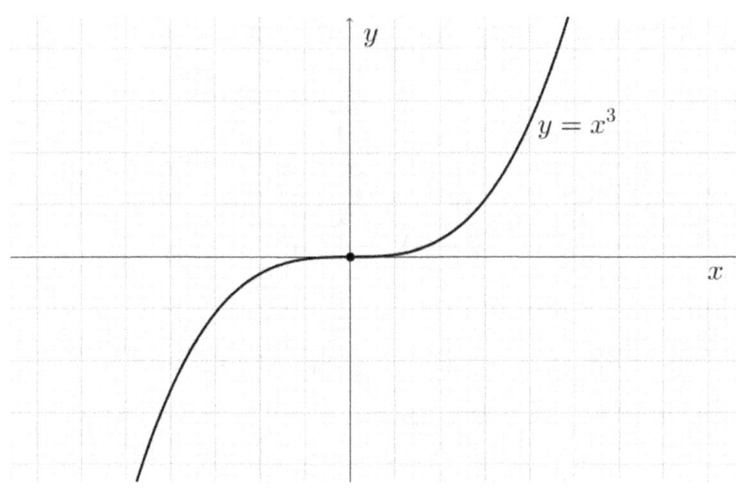

The graph of $y = x^3$ clearly shows both a stationary point
and a point of inflexion at (0,0)

Note : A point of inflexion always implies $\frac{d^2y}{dx^2}=0$

However $\frac{d^2y}{dx^2}=0$ does not always imply a point of inflexion.

For example , consider the curve $y=x^4$

$\frac{dy}{dx}=4x^3$ which implies a stationary point at the origin (0,0)

$\frac{d^2y}{dx^2}=12x^2$ which means that when $x=0$, $\frac{d^2y}{dx^2}=0$

By constructing a slope table you can see that the stationary point is a local minimum

x	-1	0	1
$\frac{dy}{dx}$	-4	0	4

A table showing $\frac{d^2y}{dx^2}$ just before and just after the point gives

x	-1	0	1
$\frac{d^2y}{dx^2}$	12	0	12

$\frac{d^2y}{dx^2}$ is not changing sign. Therefore there is no change in concavity

and $x=0$ is a not point of inflexion.

This can be clearly seen if you graph the function on a GDC

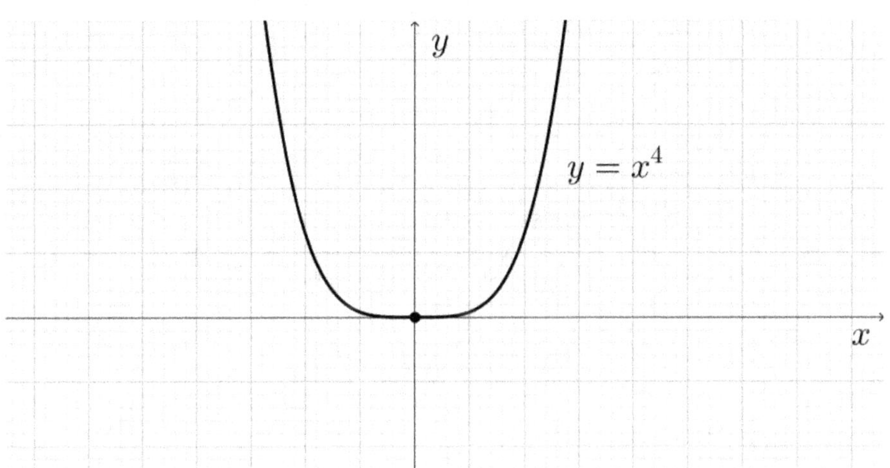

Graphing derivatives

If $f(x)=\frac{1}{3}x^3-8x^2+39x-20$ then the graph of $y=f(x)$ looks like this :

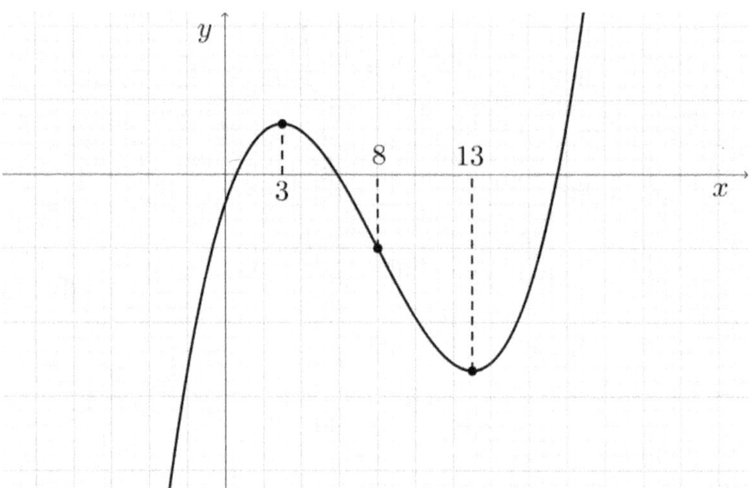

The local maximum occurs when $x=3$ and the local minimum occurs when $x=13$. The point of inflexion occurs when $x=8$.

Now $f'(x)=x^2-16x+39$ and the graph of $y=f'(x)$ looks like this :

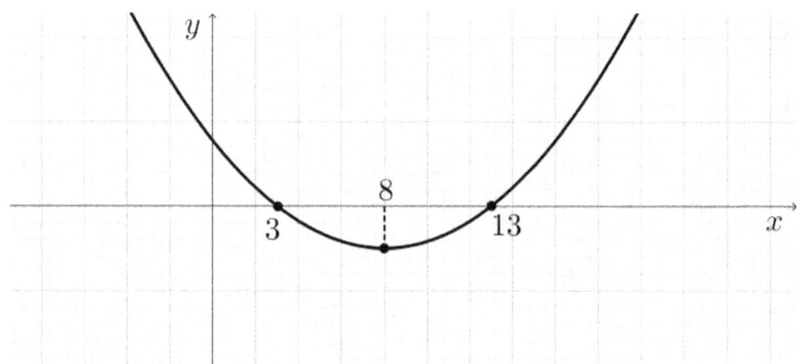

When $f'(x)=0$ we have $x=3$ and $x=13$.
This corresponds to the local maximum and the local minimum on the graph of $y=f(x)$.
The stationary point at $x=8$ corresponds to the point of inflexion on the graph of $y=f(x)$.

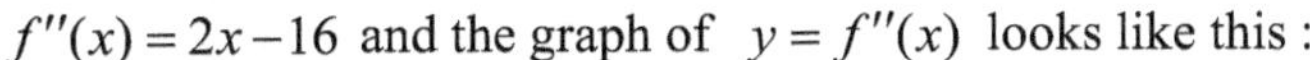

$f''(x) = 2x - 16$ and the graph of $y = f''(x)$ looks like this :

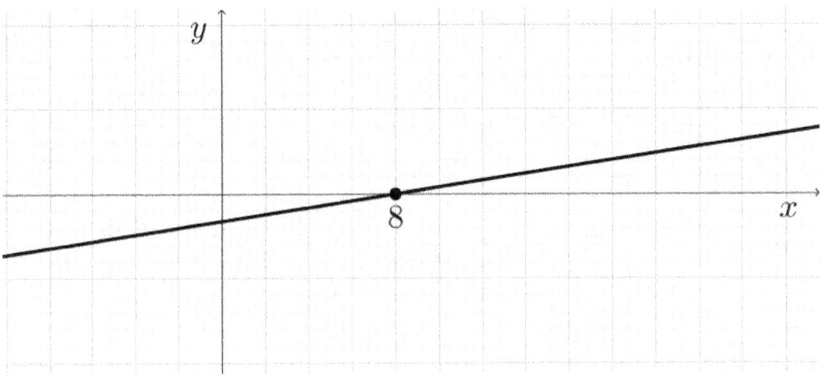

When $f''(x) = 0$ we have $x = 8$. This corresponds to the stationary point (the local minimum) on the graph of $y = f'(x)$ and the point of inflexion on the graph of $y = f(x)$.

Here are all three graphs on the same axes :

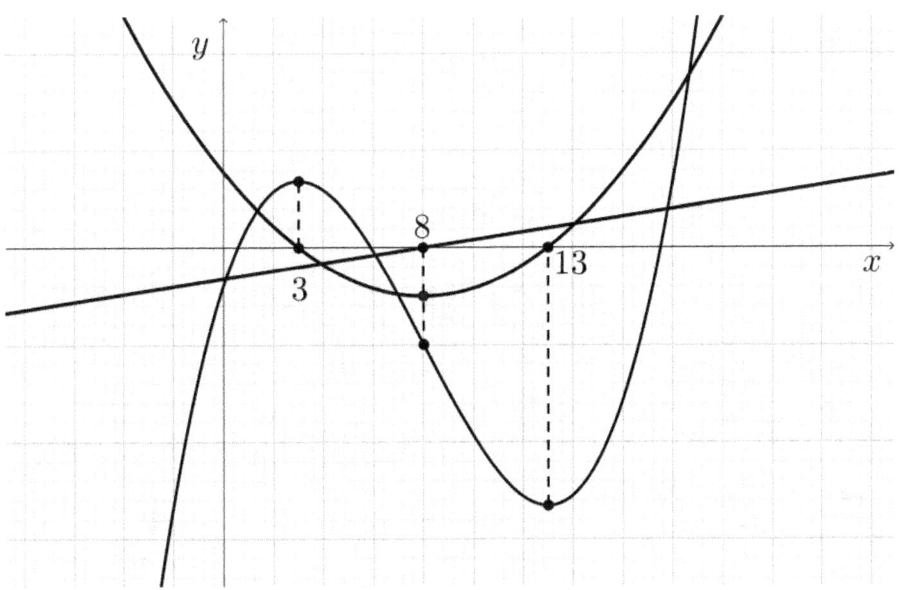

Optimization

Using differentiation to maximize or minimize a function involving two or more variables.

Example : A soda can needs to hold 300 mL of liquid.
The can has base radius r and height h .
Find an expression for the total surface area of the can in terms of r only , and hence find the dimensions of the can that result in the least amount of metal being used.
(Note : 300 mL = 300 cm^3)

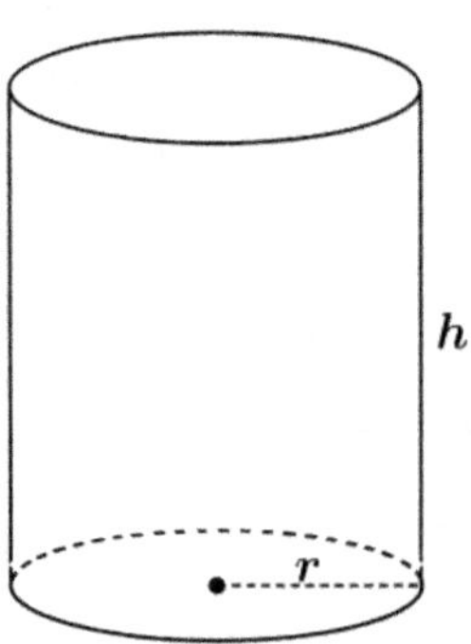

Step 1 : Find h in terms of r

Volume of a cylinder $= \pi r^2 h$

Hence $\pi r^2 h = 300$

$$h = \frac{300}{\pi r^2}$$

Step 2 : Find the total surface area of the can in terms of r :

Curved surface area $= 2\pi rh$

Top area $= \pi r^2$

Bottom area $= \pi r^2$

Hence total area $= 2\pi rh + 2\pi r^2$

$$A = 2\pi r\left(\frac{300}{\pi r^2}\right) + 2\pi r^2$$

$$= \frac{600}{r} + 2\pi r^2$$

Step 3 : Calculate $\frac{dA}{dr}$

$$A = 600r^{-1} + 2\pi r^2$$

$$\frac{dA}{dr} = -\ 600r^{-2} + 4\pi r$$

$$= -\ \frac{600}{r^2} + 4\pi r$$

Step 4 : Solve $\frac{dA}{dr} = 0$ to find the value of r

that gives the minimum value of A

$$-\ \frac{600}{r^2} + 4\pi r = 0$$

$$-\ 600 + 4\pi r^3 = 0$$

$$r^3 = \frac{600}{4\pi}$$

$$r = 3.6278 \ldots.$$

Step 5 : Calculate h

$$h = \frac{300}{\pi r^2}$$

$$= 7.2557$$

Hence the dimensions are radius = 3.63 cm and height = 7.26 cm

Example : A farmer needs to make a rectangular enclosure for his sheep. He uses an existing hedge as one of the sides of the enclosure. If he has 1000 feet of fencing available then what is the maximum area that he can enclose ?

Step 1 : Draw a diagram

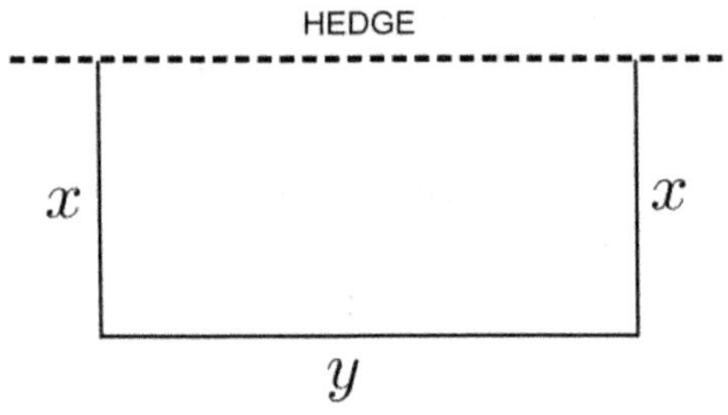

Length of fencing $= 2x + y$

Hence $2x + y = 1000$

$$y = 1000 - 2x$$

Step 2 : Find the area of the enclosure in terms of x :

Area $= xy$

$$A = x(1000 - 2x)$$
$$= 1000x - 2x^2$$

Step 3 : Calculate $\dfrac{dA}{dx}$

$$A = 1000x - 2x^2$$

$$\frac{dA}{dx} = 1000 - 4x$$

Step 4 : Solve $\frac{dA}{dx} = 0$ to find the value of x that gives the maximum value of A

$$1000 - 4x = 0$$
$$x = 250$$

Step 5 : Calculate the area

$$A = 1000x - 2x^2$$
$$A = 125000$$

Hence the maximum area is 125,000 square feet

Differentiation of $y = a\left[f(x)\right]^n$

If $y = a\left[f(x)\right]^n$ where $a, n \in \mathbb{R}$ then

$$\frac{dy}{dx} = na\left[f(x)\right]^{n-1} f'(x)$$

For example , if $y = 4\left(3x^2+5\right)^6$ then $\frac{dy}{dx} = 24\left(3x^2+5\right)^5 6x$

$$= 144x\left(3x^2+5\right)^5$$

This method follows directly from the chain rule :

Given $y = 4\left(3x^2+5\right)^6$ let $u = 3x^2+5$

This gives $y = 4u^6$

Now $\frac{dy}{dx} = \frac{dy}{du} \bullet \frac{du}{dx}$

$$= 24u^5 \bullet 6x$$

$$= 144xu^5$$

$$= 144x(3x^2+5)^5$$

Example : Given the function $y = \dfrac{4}{\sqrt{3x+1}}$ find $\dfrac{dy}{dx}$ and explain why the function has no stationary points.

Step 1 : Write function as $y = 4(3x+1)^{-\frac{1}{2}}$

Step 2 : Differentiate

$$\begin{aligned}\frac{dy}{dx} &= -2(3x+1)^{-\frac{3}{2}}(3) \\ &= \frac{-6}{\sqrt{(3x+1)^3}}\end{aligned}$$

Step 3 : Try to solve $\dfrac{dy}{dx} = 0$

$$\frac{-6}{\sqrt{(3x+1)^3}} = 0$$

$$-6 = 0$$

which means the equation has no solution.
Hence the function has no stationary points.

Product Rule

If $y = uv$ where u and v are both functions of x then

$$\frac{dy}{dx} = u\frac{dv}{dx} + v\frac{du}{dx}$$

For example , if $y = x^2(3x+1)^5$ then $u = x^2$ and $v = (3x+1)^5$

Hence $\dfrac{dy}{dx} = x^2\left[5(3x+1)^4(3)\right] + (3x+1)^5\ [2x]$

$$= 15x^2(3x+1)^4 + 2x(3x+1)^5$$

$$= x(3x+1)^4\ [15x + 2(3x+1)]$$

$$= x(3x+1)^4(21x+2)$$

Quotient Rule

If $y = \dfrac{u}{v}$ where u and v are both functions of x then

$$\frac{dy}{dx} = \frac{v\dfrac{du}{dx} - u\dfrac{dv}{dx}}{v^2}$$

For example , if $y = \dfrac{5x+1}{2x+3}$ then $u = 5x+1$ and $v = 2x+3$

Hence $\dfrac{dy}{dx} = \dfrac{(2x+3)5-(5x+1)3}{(2x+3)^2}$

$$= \frac{(10x+15)-(15x+3)}{(2x+3)^2}$$

$$= \frac{-5x+12}{(2x+3)^2}$$

Exercise 2

1) Find $\dfrac{dy}{dx}$

(a) $y=4(3x+5)^6$ (b) $y=(3x^3+4)^5$ (c) $y=8\sqrt{3x-5}$

(d) $y=\dfrac{2}{(1-4x)^2}$ (e) $y=\dfrac{1}{5x+2}$ (f) $y=\dfrac{-1}{\sqrt[3]{6x+1}}$

2) Find $f'(x)$

(a) $f(x)=x^2(6x-1)^5$ (b) $f(x)=2x\sqrt{x+1}$

(c) $f(x)=\dfrac{8x-1}{2x-1}$ (d) $f(x)=\dfrac{x}{\sqrt{2x-1}}$

3) (a) Find the gradient of the curve $y=\dfrac{1}{2x+1}$ when $x=0$

(b) Hence find the equation of the tangent to the curve when $x=0$

4) (a) Find the gradient of the curve $y=\sqrt{4x-3}$ when $x=3$

(b) Hence find the equation of the normal to the curve when $x=3$

5) Given $y=(x+1)^3(x+5)^4$

(a) Find $\dfrac{dy}{dx}$

(b) Hence find the x-coordinates of all stationary points.

6) Given that $y=\sqrt{(4x+1)^5}$ find $\dfrac{d^2y}{dx^2}$

7) Given the function $y = 2x^3 + 15x^2 - 36x + 6$

(a) Find $\frac{dy}{dx}$ and hence find the x-coordinates of all stationary points.

(b) Use a slope table to determine the nature of the stationary points.

(c) Find $\frac{d^2y}{dx^2}$.

(d) Show how you would use $\frac{d^2y}{dx^2}$ to confirm your answers to part (b).

(e) Solve $\frac{d^2y}{dx^2} = 0$ and prove that your answer represents a point of inflexion.

(f) State the values of x for which the function is increasing.

(g) State the values of x for which the function is decreasing.

(h) State the values of x for which the function is concave up.

(i) State the values of x for which the function is concave down.

8) Given the function $f(x) = 2x^3 + 15x^2 - 36x + 6$

(a) Use a graphing calculator to help draw a sketch of $y = f(x)$

(b) On the same axes sketch $y = f'(x)$ showing clearly how the two graphs relate.

(c) On the same axes sketch $y = f''(x)$ showing clearly how it relates to the previous two graphs.

9) Given the derivative function $f'(x) = x^2 - 8x + 12$

(a) Sketch $y = f'(x)$

(b) Hence sketch $y = f(x)$ given that $f(0) = 20$

10) Given the derivative function $f'(x) = -x^2 + 10x - 9$

(a) Sketch $y = f'(x)$

(b) Hence sketch $y = f(x)$ given that $f(0) = 18$

11) A cuboid box has length three times its width and height h.
If the volume needs to be 2304 cm^3 then calculate smallest possible surface area of the box.

12) A rectangular piece of cardboard has dimensions 12 cm by 10 cm.
Small squares of side x are cut from the four corners of the cardboard.
The resulting four flaps are folded up to form an open box of depth x.

(a) Express the volume , V, of the box in terms of $\boldsymbol{x}$

(b) Find $\dfrac{dV}{dx}$

(c) Hence calculate the value of x that gives the maximum possible volume of the box.

13) A right circular cone has base radius r and height h.
The curved surface area of the cone is 10π cm^2.

(a) Find an expression for the volume of the cone in terms of r only.

(b) Hence find the **exact** value of the radius that gives the maximum possible volume.

14) (a) Find the x-coordinate of the stationary point on the function

$$y = 2x - 9x^{\frac{2}{3}} \text{ where } x > 0$$

(b) Prove that it is a minimum point.

15) (a) Sketch the parabola $y = 9 - x^2$

(b) The point $P(a\,,b)$ lies on the parabola.
The point O is the origin (0 ,0)
Find an expression for the distance OP in terms of b only.

(c) Hence find the shortest possible distance of P from the origin.
Give your answer correct to 3 decimal places.

Implicit Differentiation

We know $\frac{d}{dx}\left(3x^4\right) = 12x^3$ and $\frac{d}{dt}\left(3t^4\right) = 12t^3$

But how do we find $\frac{d}{dx}\left(3t^4\right)$?

This requires what we call "implicit differentiation".

$$\frac{d}{dx}\left(3t^4\right) = \frac{dt}{dx}\frac{d}{dt}\left(3t^4\right)$$

$$= \frac{dt}{dx}12t^3$$

Here the differential $\frac{dt}{dx}$ forms part of the answer.

For example if $3x^4 + 3t^4 = 1$ and you need to find $\frac{dt}{dx}$ then do the following :

$$\frac{d}{dx}\left(3x^4 + 3t^4\right) = \frac{d}{dx}(1)$$

$$\frac{d}{dx}\left(3x^4\right) + \frac{d}{dx}\left(3t^4\right) = 0$$

$$12x^3 + \frac{dt}{dx}12t^3 = 0$$

$$\frac{dt}{dx}12t^3 = -12x^3$$

$$\frac{dt}{dx} = \frac{-12x^3}{12t^3}$$

$$= \frac{-x^3}{t^3}$$

This result could also have been obtained using a previous method :

$$3x^4 + 3t^4 = 1$$

$$3t^4 = 1 - 3x^4$$

$$t^4 = \frac{1}{3}\left(1 - 3x^4\right)$$

$$t = \left[\frac{1}{3}\left(1 - 3x^4\right)\right]^{\frac{1}{4}}$$

$$\frac{dt}{dx} = \frac{1}{4}\left[\frac{1}{3}\left(1 - 3x^4\right)\right]^{-\frac{3}{4}}\left(-4x^3\right)$$

$$= -x^3\left[\frac{1}{3}\left(1 - 3x^4\right)\right]^{-\frac{3}{4}}$$

$$= -x^3\left[t^4\right]^{-\frac{3}{4}}$$

$$= \frac{-x^3}{t^3}$$

But clearly this involves a lot more work , and sometimes it is not possible to express x in terms of t .

Example : Given the circle $x^2 + y^2 = 13$ find $\frac{dy}{dx}$ and hence find the equation of the tangent to the circle at the point (2, 3)

Step 1 : Differentiate both sides with respect to x

(in other words find $\frac{d}{dx}$ of both sides)

$$\frac{d}{dx}\left(x^2 + y^2\right) = \frac{d}{dx}(13)$$

$$\frac{d}{dx}\left(x^2\right) + \frac{d}{dx}\left(y^2\right) = 0$$

$$2x + \frac{dy}{dx}\frac{d}{dy}\left(y^2\right) = 0$$

$$2x + \frac{dy}{dx}2y = 0$$

$$\frac{dy}{dx} = \frac{-2x}{2y}$$

$$\frac{dy}{dx} = \frac{-x}{y}$$

Step 2 : Substitute $x = 2$ and $y = 3$ to find the gradient of the tangent

$$\frac{dy}{dx} = \frac{-2}{3}$$

Step 3 : Use $y = mx + c$ to find the equation

$$y = -\frac{2}{3}x + c$$

At (2, 3) we have $3 = -\frac{2}{3}(2) + c$

$$c = \frac{13}{3}$$

Hence $y = -\frac{2}{3}x + \frac{13}{3}$

Short-cut :

There is no need to show all of the steps when doing implicit differentiation.

$\frac{d}{dx}\left(x^2 + y^2\right) = \frac{d}{dx}(13) \quad \Rightarrow \quad 2x + 2y\frac{dy}{dx} = 0$ is sufficient

Note : when finding $\frac{d}{dx}\left(y^2\right)$ differentiate "as normal"

and then just put $\frac{dy}{dx}$ at the end.

For example : $\frac{d}{dx}\left(5y^3\right) = 15y^2\frac{dy}{dx}$

Using the product rule with implicit differentiation :

To find $\frac{d}{dx}\left(x^2 y^3\right)$ treat this as a product uv.

We then get $x^2 \frac{d}{dx}\left(y^3\right) + y^3 \frac{d}{dx}\left(x^2\right)$

This then becomes $x^2\ 3y^2 \frac{dy}{dx} + y^3\ 2x$

(This is the equivalent of $u\frac{dv}{dx}+v\frac{du}{dx}$)

Example :

Given $4x^2+3xy+6y^2=7$ find $\frac{dy}{dx}$ and hence find the gradients at the point where $x=1$

Step 1 : Differentiate both sides with respect to x

$$\frac{d}{dx}\left(4x^2+3xy+6y^2\right)=\frac{d}{dx}(7)$$

$$8x+3x\frac{d}{dx}(y)+y\frac{d}{dx}(3x)+12y\frac{dy}{dx} = 0$$

Note : The two center terms are from the product rule and the method is shown in more detail than is normally required.

$$8x+3x(1)\frac{dy}{dx}+y(3)+12y\frac{dy}{dx} = 0$$

$$8x+3x\frac{dy}{dx}+3y+12y\frac{dy}{dx} = 0$$

$$3x\frac{dy}{dx}+12y\frac{dy}{dx} = -8x-3y$$

$$\frac{dy}{dx}(3x+12y) = -8x-3y$$

$$\frac{dy}{dx} = \frac{-8x-3y}{3x+12y}$$

Step 2 : Find y when $x=1$

$$4x^2+3xy+6y^2=7 \Rightarrow 4+3y+6y^2=7$$

$$6y^2+3y-3=0$$

$$2y^2+y-1=0$$

$$(2y-1)(y+1)=0$$

$$y=\frac{1}{2},-1$$

Step 3 : Substitute into $\frac{dy}{dx} = \frac{-8x-3y}{3x+12y}$ to find the gradient

At $\left(1\,,\,\frac{1}{2}\right)$ we have $\frac{dy}{dx} = \frac{-8-\frac{3}{2}}{3+6}$

$$= \frac{-19}{18}$$

At $(1\,,-1)$ we have $\frac{dy}{dx} = \frac{-8+3}{3-12}$

$$= \frac{5}{9}$$

Example :

Given $3x^2 - y^2 = 6$ use implicit differentiation to find $\dfrac{d^2y}{dx^2}$

Step 1 : Differentiate both sides with respect to x

$$\frac{d}{dx}\left(3x^2 - y^2\right) = \frac{d}{dx}(6)$$

$$6x - 2y\frac{dy}{dx} = 0$$

$$\frac{dy}{dx} = \frac{6x}{2y}$$

$$= \frac{3x}{y}$$

Step 2 : Differentiate both sides again with respect to x

Start from $\quad 6x - 2y\dfrac{dy}{dx} = 0$

$$\frac{d}{dx}\left(6x - 2y\frac{dy}{dx}\right) = \frac{d}{dx}(0)$$

$$6 - \left[(2y)\left(\frac{d^2y}{dx^2}\right) + \left(\frac{dy}{dx}\right)\left(2\frac{dy}{dx}\right)\right] = 0$$

(Note : The last two terms are from the product rule)

$$6 - 2y\frac{d^2y}{dx^2} - 2\left(\frac{dy}{dx}\right)^2 = 0$$

Step 3 : Replace $\frac{dy}{dx}$ by $\frac{3x}{y}$

$$6-2y\frac{d^2y}{dx^2}-2\left(\frac{3x}{y}\right)^2 = 0$$

$$6-2y\frac{d^2y}{dx^2}-\frac{18x^2}{y^2} = 0$$

$$2y\frac{d^2y}{dx^2} = 6-\frac{18x^2}{y^2}$$

$$\frac{d^2y}{dx^2} = \frac{3}{y}-\frac{9x^2}{y^3}$$

$$= \frac{3y^2-9x^2}{y^3}$$

Example :

Given $3x - x^3 y + y^2 = 1$ find $\dfrac{dy}{dx}$ and $\dfrac{d^2y}{dx^2}$ at the point (0,1)

Step 1 : Differentiate both sides with respect to x

$$\frac{d}{dx}\left(3x - x^3 y + y^2\right) = \frac{d}{dx}(1)$$

$$3 - \left[x^3\left(1\frac{dy}{dx}\right) + y\left(3x^2\right)\right] + 2y\frac{dy}{dx} = 0$$

$$3 - x^3\frac{dy}{dx} - 3x^2 y + 2y\frac{dy}{dx} = 0$$

Step 2 : Substitute $x = 0$ and $y = 1$ to find $\dfrac{dy}{dx}$

$$3 + 2\frac{dy}{dx} = 0 \Rightarrow \frac{dy}{dx} = \frac{-3}{2}$$

Step 3 : Differentiate both sides again with respect to x

Start from the end of Step 1

$$3 - x^3\frac{dy}{dx} - 3x^2 y + 2y\frac{dy}{dx} = 0$$

$$\frac{d}{dx}\left(3 - x^3\frac{dy}{dx} - 3x^2 y + 2y\frac{dy}{dx}\right) = \frac{d}{dx}(0)$$

$$-\left[x^3\frac{d^2y}{dx^2} + \frac{dy}{dx}3x^2\right] - \left[3x^2\frac{dy}{dx} + y(6x)\right] + \left[2y\frac{d^2y}{dx^2} + \frac{dy}{dx}\left(2\frac{dy}{dx}\right)\right] = 0$$

$$-x^3\frac{d^2y}{dx^2} - \frac{dy}{dx}3x^2 - 3x^2\frac{dy}{dx} - 6xy + 2y\frac{d^2y}{dx^2} + 2\left(\frac{dy}{dx}\right)^2 = 0$$

Step 4 : Substitute $x=0$ $y=1$ and $\frac{dy}{dx}=\frac{-3}{2}$ to find $\frac{d^2y}{dx^2}$

$$2\frac{d^2y}{dx^2}+2\left(\frac{-3}{2}\right)^2 = 0$$

$$\frac{d^2y}{dx^2} = -\frac{9}{4}$$

Example :

A ladder 13 feet in length is leaning against a vertical wall.
The top of the ladder is sliding down the wall at a rate of 2 feet per second.
Calculate the speed at which the foot of the ladder is moving away from the wall when the foot of the ladder is 12 feet from the wall.

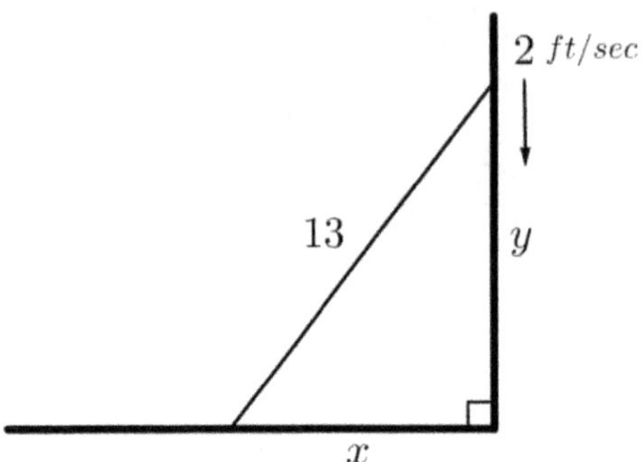

Step 1 : Write $x^2 + y^2 = 13^2$

Step 2 : Differentiate both sides with respect to *t (implicitly)*

$$\frac{d}{dt}\left(x^2 + y^2\right) = \frac{d}{dt}\left(13^2\right)$$

$$2x\frac{dx}{dt} + 2y\frac{dy}{dt} = 0$$

Step 3 : Substitute $x = 12$, $y = 5$ and $\frac{dy}{dt} = 2$

$$24\frac{dx}{dt} + 10(2) = 0$$

$$\frac{dx}{dt} = \frac{-20}{24}$$

$$= -\frac{5}{6}$$

Hence the foot of the ladder is moving at $\frac{5}{6}$ ft/sec away from the base of the wall.

Differentiation of trigonometric functions

y	$\frac{dy}{dx}$
$\sin x$	$\cos x$
$\cos x$	$-\sin x$
$\tan x$	$\sec^2 x$
$\operatorname{cosec} x$	$-\operatorname{cosec} x \cot x$
$\sec x$	$\sec x \tan x$
$\cot x$	$-\operatorname{cosec}^2 x$

Make sure you use the chain rule when differentiating trig functions.

For example : $y = \sin 4x \Rightarrow \frac{dy}{dx} = 4\cos 4x$

$$y = \tan\left(3x^2\right) \Rightarrow \frac{dy}{dx} = 6x\sec^2\left(3x^2\right)$$

$$y = \operatorname{cosec}\,(5x+\pi) \Rightarrow \frac{dy}{dx} = -5\,\operatorname{cosec}\,(5x+\pi)\cot(5x+\pi)$$

$$y = \sin^3 4x \Rightarrow y = (\sin 4x)^3 \quad \Rightarrow \frac{dy}{dx} = 3(\sin 4x)^2\, 4\cos 4x$$

$$= 12\sin^2 4x\cos 4x$$

Important : For differentiation , all trig functions MUST be in radians.

Differentiation of $y = \ln x$ and $y = e^x$

y	$\frac{dy}{dx}$
$\ln x$	$\frac{1}{x}$
e^x	e^x

Again , make sure you use the chain rule when differentiating these functions.

$$y = \ln(4x+3) \Rightarrow \frac{dy}{dx} = 4\left(\frac{1}{4x+3}\right) = \frac{4}{4x+3}$$

$$y = \ln(x^2+3x+4) \Rightarrow \frac{dy}{dx} = \frac{2x+3}{x^2+3x+4}$$

$$y = e^{5x} \Rightarrow \frac{dy}{dx} = 5e^{5x}$$

$$y = e^{3x^2+7x} \Rightarrow \frac{dy}{dx} = (6x+7)e^{3x^2+7x}$$

Why is $e = 2.71828$ ….. special ?

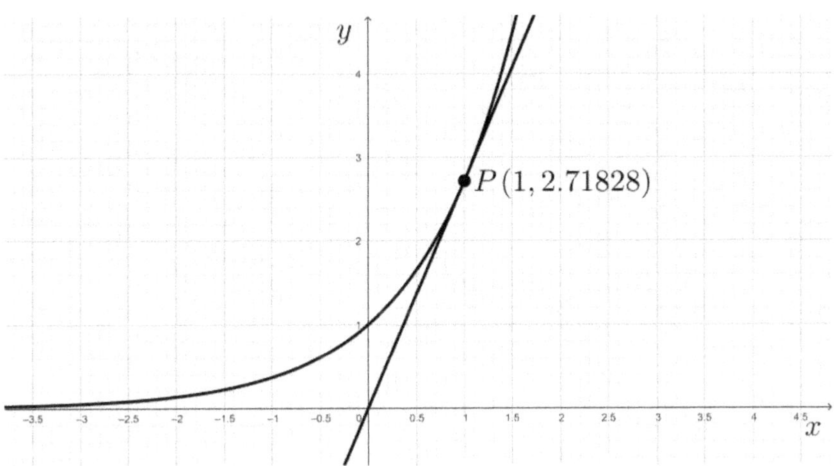

The graph of $y = e^x$ is shown above.

It is an exponential function so it rapidly increases as x increases.

The x-axis is an asymptote to the curve.

The tangent to the curve when $x = 1$ is also shown.

This tangent has gradient 2.71828 which is the same as the y-coordinate of P.

This means that y and $\dfrac{dy}{dx}$ both have the same value at P.

In fact , if you take any point on the curve , you will find that the y-coordinate and the value of $\dfrac{dy}{dx}$ at that point are the same. This is the unique property of e.

Hence $y = e^x \Rightarrow \dfrac{dy}{dx} = e^x$

Differentiation of $y = \log_a x$ and $y = a^x$

To differentiate $y = \log_a x$ you need to change the base.

$$y = \log_a x \Rightarrow a^y = x$$

$$\Rightarrow \ln a^y = \ln x$$

$$\Rightarrow y \ln a = \ln x$$

$$\Rightarrow y = \left(\frac{1}{\ln a}\right)\ln x$$

$$\Rightarrow \frac{dy}{dx} = \left(\frac{1}{\ln a}\right)\frac{1}{x}$$

$$\frac{dy}{dx} = \frac{1}{x \ln a}$$

You do not need to show all of these steps. It is acceptable to write

$$y = \log_a x \quad \Rightarrow \frac{dy}{dx} = \frac{1}{x \ln a}$$

Example : Differentiate (a) $y = \log_4 x$ (b) $y = \log(2x+1)$

(a) $y = \log_4 x \Rightarrow \dfrac{dy}{dx} = \dfrac{1}{x \ln 4}$

(b) $y = \log(2x+1) \Rightarrow \dfrac{dy}{dx} = \dfrac{2}{(2x+1)\ln 10}$

Note : $\log x = \log_{10} x$

To differentiate $y = a^x$ you need to convert from a to e

$$y = a^x \quad \Rightarrow \quad \ln y = \ln a^x$$

$$\Rightarrow \quad \ln y = x \ln a$$

$$\Rightarrow \quad y = e^{x \ln a}$$

$$\Rightarrow \quad \frac{dy}{dx} = (\ln a) e^{x \ln a}$$

$$\Rightarrow \quad \frac{dy}{dx} = (\ln a) y$$

$$\Rightarrow \quad \frac{dy}{dx} = (\ln a) a^x$$

Again , you do not need to show all of these steps. It is acceptable to write

$$y = a^x \Rightarrow \frac{dy}{dx} = (\ln a) a^x$$

y	$\frac{dy}{dx}$
$\log_a x$	$\frac{1}{x \ln a}$
a^x	$(\ln a) a^x$

Example : Given $y^2 = e^{\sin x+\cos y}$ find $\dfrac{dy}{dx}$

Step 1 : Differentiate both sides with respect to x

$$\frac{d}{dx}\left(y^2\right) = \frac{d}{dx}\left(e^{\sin x+\cos y}\right)$$

$$2y\frac{dy}{dx} = \left(e^{\sin x+\cos y}\right)\left(\cos x - \sin y\frac{dy}{dx}\right)$$

Step 2 : Expand and rearrange to isolate $\dfrac{dy}{dx}$

$$2y\frac{dy}{dx} = \left(e^{\sin x+\cos y}\right)\cos x - \left(e^{\sin x+\cos y}\right)\sin y\frac{dy}{dx}$$

$$2y\frac{dy}{dx} + \left(e^{\sin x+\cos y}\right)\sin y\frac{dy}{dx} = \left(e^{\sin x+\cos y}\right)\cos x$$

$$\frac{dy}{dx}\left[2y + \left(e^{\sin x+\cos y}\right)\sin y\right] = \left(e^{\sin x+\cos y}\right)\cos x$$

$$\frac{dy}{dx} = \frac{\left(e^{\sin x+\cos y}\right)\cos x}{2y + \left(e^{\sin x+\cos y}\right)\sin y}$$

Differentiation of inverse trigonometric functions

Consider $y = \arcsin x$ (or $y = \sin^{-1} x$)

$$y = \arcsin x \quad \Rightarrow \quad x = \sin y$$

$$\Rightarrow \quad \frac{dx}{dy} = \cos y$$

$$\Rightarrow \quad \frac{dy}{dx} = \frac{1}{\cos y}$$

$$= \frac{1}{\sqrt{\cos^2 y}}$$

$$= \frac{1}{\sqrt{1-\sin^2 y}}$$

$$= \frac{1}{\sqrt{1-x^2}}$$

Similarly $y = \arccos x \Rightarrow \dfrac{dx}{dy} = \dfrac{-1}{\sqrt{1-x^2}}$

If $y = \arctan x$ then $x = \tan y$

$$\Rightarrow \quad \frac{dx}{dy} = \sec^2 y$$

$$\Rightarrow \quad \frac{dy}{dx} = \frac{1}{\sec^2 y}$$

$$= \frac{1}{1+\tan^2 y}$$

$$= \frac{1}{1+x^2}$$

y	$\frac{dy}{dx}$
$\arcsin x$	$\frac{1}{\sqrt{1-x^2}}$
$\arccos x$	$\frac{-1}{\sqrt{1-x^2}}$
$\arctan x$	$\frac{1}{1+x^2}$

Example : Given $y = \arctan(3x^2 + 1)$ find $\frac{dy}{dx}$

Step 1 : Let $u = 3x^2 + 1$

Then $\frac{du}{dx} = 6x$ and $y = \arctan u$

Step 2 : Differentiate $y = \arctan u$
(you can give the answer without any working)

$$\frac{dy}{du} = \frac{1}{1+u^2}$$

Step 3 : Use the Chain Rule to get the required answer

$$\frac{dy}{dx} = \frac{dy}{du} \bullet \frac{du}{dx}$$

$$= \frac{1}{1+u^2} \bullet 6x$$

$$= \frac{6x}{1+\left(3x^2+1\right)^2}$$

$$= \frac{6x}{9x^4+6x^2+2}$$

Example :

Find the exact value of the gradient of the tangent to $y = x\arcsin(3x)$ at the point where $x = \frac{1}{6}$

Step 1 : Find $\frac{dy}{dx}$

$$y = x\arcsin(3x)$$

$$\frac{dy}{dx} = x\frac{3}{\sqrt{1-(3x)^2}} + \arcsin(3x)$$

$$= \frac{3x}{\sqrt{1-9x^2}} + \arcsin(3x)$$

Step 2 : Substitute $x = \frac{1}{6}$ to find the gradient of the tangent

$$\frac{dy}{dx} = \frac{\frac{1}{2}}{\sqrt{1-\frac{9}{36}}} + \arcsin\left(\frac{1}{2}\right)$$

$$= \frac{3}{\sqrt{27}} + \frac{\pi}{6}$$

$$= \frac{3}{3\sqrt{3}} + \frac{\pi}{6}$$

$$= \frac{\sqrt{3}}{3} + \frac{\pi}{6}$$

Exercise 3

1) Find $\frac{dy}{dx}$

(a) $y = e^{2x} \sin 2x$ (b) $y = \ln\left(x^2 + 5x + 1\right)$ (c) $y = \sec(4x - \pi)$

(d) $y = \cos\left(4 - x^2\right)$ (e) $y = \frac{x}{\ln x}$ (f) $y = \log_4 x$

(g) $y = 4^x$ (h) $y = \ln(\sec 2x)$ (i) $y = \text{cosec}^2(1 - x)$

2) Find $f'(x)$

(a) $f(x) = e^{\tan 2x}$ (b) $f(x) = \sin^3 4x$

(c) $f(x) = \text{cosec}^2 3x$ (d) $f(x) = \frac{1}{\sqrt{\cot 4x}}$

3) Use the quotient rule and $\tan x = \frac{\sin x}{\cos x}$ to show that

$$\frac{d}{dx}(\tan x) = \sec^2 x$$

4) Given $x^2 + xy + y^2 = 3$

(a) Find an expression for $\frac{dy}{dx}$ in terms of x and y

(b) Hence find $\frac{dy}{dx}$ at the two points where $x = 1$

(c) Find $\frac{d^2y}{dx^2}$ at the point $(1,1)$

5) Differentiate (a) $y = \cos^{-1}\left(1 - x^2\right)$ (b) $y = \sin^{-1}(5x)$

(c) $y = \tan^{-1}\left(e^x\right)$

6) If $y = \tan^{-1}(\sin x)$ show that $\dfrac{d^2y}{dx^2} = \dfrac{\sin^3 x - 3\sin x}{\left(1 + \sin^2 x\right)^2}$

7) Given $y = \sec 3x$ find $\dfrac{d^2y}{dx^2}$

8) Given $y = (\ln x)^3$ find $\dfrac{d^2y}{dx^2}$

9) Given $x^3y^2 + x + y = 7$ find $\dfrac{dy}{dx}$

10) Given $y^2 = \sin(xy)$ find $\dfrac{dy}{dx}$

11) Let $y = \cos^2 x$ for $0 \le x \le \pi$

(a) Find $\dfrac{dy}{dx}$

(b) Solve $\dfrac{dy}{dx} = 2y$

12) Find the gradient of the normal to the curve $xy^2 = \cos(\pi y)$ at the point $(-1,1)$

13) Find the **exact** coordinates of the maximum point on the curve $y = \dfrac{x}{e^{2x}}$ and justify that it is a maximum point.

14) A right triangle has hypotenuse of fixed length 20 cm and legs of variable lengths x and y. If x is decreasing at 4 cm/sec then find the corresponding rate of increase in y when $x = 12$ cm.

15) A rocket is rising vertically from a launch site. When it is at a height of 500 meters the speed of the rocket is 200 meters per second. An observer is located 1200 meters from the base of the launch tower.

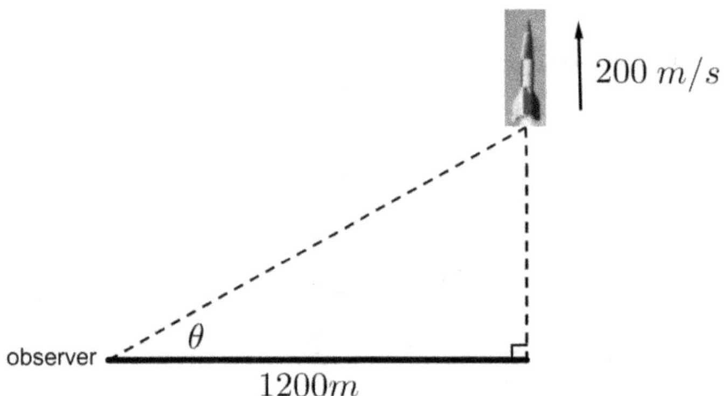

(a) Calculate the rate of change of the distance between the observer and the rocket at that time.

(b) Calculate the corresponding rate of change in the angle of elevation between the observer and the rocket. (Assume the angle is measured from ground level)

L'Hopitals Rule

This is used when $\lim\limits_{x\to a} \dfrac{f(x)}{g(x)}$ gives the result $\dfrac{0}{0}$ or $\dfrac{\infty}{\infty}$

In which case the following is true :

$$\lim_{x\to a} \frac{f(x)}{g(x)} = \lim_{x\to a} \frac{f'(x)}{g'(x)}$$

For example , if we need to find $\lim\limits_{x\to 0} \dfrac{\sin x}{x}$ we first substitute $x=0$

and see what happens. Here we get $\dfrac{\sin 0}{0}$ which gives the result $\dfrac{0}{0}$.

This means we can apply L'Hopitals Rule :

Separately differentiating the numerator and the denominator gives

$$\lim_{x\to 0} \frac{\sin x}{x} = \lim_{x\to 0} \frac{\cos x}{1}$$

Substitute $x=0$ again to give $\dfrac{1}{1}$. Hence $\lim\limits_{x\to 0} \dfrac{\sin x}{x} = 1$

This can be checked with a calculator. Give x a random very small value and see what happens :

$$\frac{\sin 0.1}{0.1} = \frac{0.0998334}{0.1}$$

$$= 0.998$$

$$\simeq 1$$

(Don't forget that you need to use radians here)

Example : (a) Explain why it is appropriate to use L'Hopitals Rule

to find $\lim\limits_{x\to\infty} \dfrac{12x^2+6x}{3x^2+1}$

(b) Find $\lim\limits_{x\to\infty} \dfrac{12x^2+6x}{3x^2+1}$

Step 1 : Substitute $x=\infty$

$$\frac{12x^2+6x}{3x^2+1} \quad \to \quad \frac{\infty}{\infty}$$

This result satisfies the condition for L'Hopitals Rule

Step 2 : Differentiate the numerator and denominator separately.

$$\lim_{x\to\infty} \frac{12x^2+6x}{3x^2+1} \quad \to \quad \lim_{x\to\infty} \frac{24x+6}{6x}$$

Step 3 : Substitute $x=\infty$ again

$$\frac{24x+6}{6x} \quad \to \quad \frac{\infty}{\infty}$$

Step 4 : Repeat the L'Hopitals Rule process

$$\lim_{x\to\infty} \frac{24x+6}{6x} \quad \to \quad \lim_{x\to\infty} \frac{24}{6}$$

$$\to \quad \lim_{x\to\infty} 4$$

$$= 4$$

Note : L'Hopitals Rule can be repeated many times on a given function (as long as each time the limit gives the result $\dfrac{\infty}{\infty}$ or $\dfrac{0}{0}$)

Example :

Use L'Hopitals Rule to find $\lim\limits_{x \to 1} \dfrac{\ln x}{\sin 4\pi x}$

Step 1 : Substitute $x = 1$

$$\frac{\ln x}{\sin 4\pi x} \rightarrow \frac{\ln 1}{\sin 4\pi}$$

$$= \frac{0}{0}$$

Step 2 : Differentiate the numerator and denominator separately.

$$\lim_{x \to 1} \frac{\ln x}{\sin 4\pi x} = \lim_{x \to 1} \frac{\frac{1}{x}}{4\pi \cos 4\pi x}$$

Step 3 : Substitute $x = 1$ again

$$\frac{\frac{1}{x}}{4\pi \cos 4\pi x} = \frac{\frac{1}{1}}{4\pi \cos 4\pi}$$

$$= \frac{1}{4\pi}$$

Exercise 4

1) Use L'Hopital's Rule to find

(a) $\lim_{x\to 0} \dfrac{\tan 3x}{\sin 4x}$ (b) $\lim_{x\to \infty} \dfrac{x^2+x}{2x^2+1}$ (c) $\lim_{x\to 2} \dfrac{\ln\left(\frac{x}{2}\right)}{x^2-4}$

2) Explain why L'Hopital's Rule cannot be used to find $\lim_{x\to 0} \dfrac{\cos x}{x}$

3) (a) Use a GDC to graph $f(x)=\dfrac{\ln(\cos x)}{\sin(x^2)}$ for $-3 \le x \le 3$ and $-1 \le y \le 1$

(b) What happens when you try to find $f(0)$ by using CALC and VALUE on your GDC ? Explain.
[Note : these are functions on the TI-84 Plus calculator.
Other models may be different].

(c) Use CALC and VALUE to find (i) $f(0.1)$ (ii) $f(-0.1)$

(d) What does this indicate about $\lim_{x\to 0} \dfrac{\ln(\cos x)}{\sin(x^2)}$

(e) Confirm your answer to part (d) by using L'Hopital's Rule.

4) When possible , find the following limits using L'Hopital's Rule.
Give a reason if using L'Hopital's Rule is not possible.

(a) $\lim_{x\to 0} \dfrac{x}{1-\cos x}$ (b) $\lim_{x\to \infty} \dfrac{5x^2+2x}{2x+1}$ (c) $\lim_{x\to e} \dfrac{1-\ln x}{x-e}$

5) Use L'Hopital's Rule to find (a) $\lim_{x\to 0} \dfrac{\cos 2x-\cos x}{\sin 2x-\sin x}$

(b) $\lim_{x\to 0} \dfrac{\tan 2x-\tan x}{\tan 2x+\sin x}$

Integration

Integration is the reverse of differentiation.

The integral of $3x^2$ is $x^3 + c$ where c is any constant.

This is because $\frac{d}{dx}\left(x^3 + c\right) = 3x^2$

If you know , for example , that $x^3 + c = 14$ when $x = 2$ then you can find the value of c. In this case it would be 6.

The notation for integration is $\int f(x)\ dx$

For example : $\int 3x^2\ dx = x^3 + c$

$$\int 10x^4\ dx = 2x^5 + c$$

You can use the following formula to quickly integrate polynomial functions

$$\int ax^n\ dx = \frac{ax^{n+1}}{n+1} + c$$

For example : $\int 2x^7\ dx = \frac{2x^8}{8} + c$

$$= \frac{x^8}{4} + c$$

$$\int 4x + 3\ dx = \frac{4x^2}{2} + 3x + c$$

$$= 2x^2 + 3x + c$$

$$\int 12x^3 + 6x^2 + 8x + 5 \; dx = \frac{12x^4}{4} + \frac{6x^3}{3} + \frac{8x^2}{2} + 5x + c$$

$$= 3x^4 + 2x^3 + 4x^2 + 5x + c$$

$$\int \frac{4}{x^3} \; dx = \int 4x^{-3} \; dx$$

$$= \frac{4x^{-2}}{-2} + c$$

$$= -\frac{2}{x^2} + c$$

$$\int 12 \sqrt[3]{x} \; dx = \int 12\, x^{\frac{1}{3}} \; dx$$

$$= \frac{12\, x^{\frac{4}{3}}}{\frac{4}{3}} + c$$

$$= 9\, x^{\frac{4}{3}} + c$$

$$= 9 \sqrt[3]{x^4} + c$$

$$\int \frac{4}{\sqrt{x}} \; dx = \int 4x^{-\frac{1}{2}} \; dx$$

$$= \frac{4x^{\frac{1}{2}}}{\frac{1}{2}} + c$$

$$= 8\sqrt{x} + c$$

Integration by inspection

For example the integral of $\cos x$ is $\sin x$ (plus some constant) because you know that $\dfrac{d}{dx}(\sin x) = \cos x$

Likewise $\displaystyle\int \frac{1}{x}\, dx = \ln x + c$ because $\dfrac{d}{dx}(\ln x) = \dfrac{1}{x}$

It is usually necessary to adjust the answer to take into account the chain rule.

For example $\displaystyle\int \cos 3x\, dx = \frac{1}{3}\sin 3x + c$ because $\dfrac{d}{dx}(\sin 3x) = 3\cos 3x$

It is always a good idea to check your work when using this method.
You should differentiate your answer and make sure it goes back to the original integral.

To find $\displaystyle\int \sin^3 2x \cos 2x\, dx$ you first need to notice that $\cos 2x$ is the differential of $\sin 2x$ (apart from the adjustment required by the chain rule).
Then consider $y = \sin^4 2x$ (note that the power is one more than the power of *sin* in the integral)

$$y = \sin^4 2x \quad \Rightarrow \quad y = (\sin 2x)^4$$

$$\Rightarrow \quad \frac{dy}{dx} = 4(\sin 2x)^3(2\cos 2x)$$

$$= 8\sin^3 2x \cos 2x$$

This is the same as the integral except for the 8

Hence $y = \dfrac{1}{8}\sin^4 2x \;\Rightarrow\; \dfrac{dy}{dx} = \sin^3 2x \cos 2x$

Therefore $\displaystyle\int \sin^3 2x \cos 2x\, dx = \frac{1}{8}\sin^4 2x + c$

Check : Let $y = \frac{1}{8}\sin^4 2x$

$$= \frac{1}{8}(\sin 2x)^4$$

Then $$\frac{dy}{dx} = \frac{1}{8}\left[4(\sin 2x)^3\, 2\cos 2x\right]$$

$$= \sin^3 2x \cos 2x$$

To find $\int \frac{x}{3x^2+1}\; dx$ you need to notice that the numerator is the differential of the denominator (apart from the adjustment required by the chain rule). This implies that $y = \ln\left(3x^2+1\right)$ is a good place to start when figuring out the answer.

$$y = \ln\left(3x^2+1\right) \quad \Rightarrow \quad \frac{dy}{dx} = \frac{6x}{3x^2+1}$$

Hence $$y = \frac{1}{6}\ln\left(3x^2+1\right) \quad \Rightarrow \quad \frac{dy}{dx} = \frac{x}{3x^2+1}$$

Therefore $$\int \frac{x}{3x^2+1}\; dx = \frac{1}{6}\ln\left(3x^2+1\right) + c$$

Check : Let $y = \frac{1}{6}\ln\left(3x^2+1\right)$

Then $$\frac{dy}{dx} = \frac{1}{6}\left[\frac{6x}{3x^2+1}\right]$$

$$= \frac{x}{3x^2+1}$$

When solving $\int xe^{4x^2+3}\ dx$ we can again see that x is the differential of $4x^2+3$ (apart from the adjustment required by the chain rule).

Hence $\int xe^{4x^2+3}\ dx \ = \ \frac{1}{8}e^{4x^2+3} \ + \text{c}$

Check : Let $y = \frac{1}{8}e^{4x^2+3}$

Then $\frac{dy}{dx} \ = \ \frac{1}{8}\left[e^{4x^2+3}(8x)\right]$

$= \ xe^{4x^2+3}$

Whenever you see an integral involving a product of two functions you should always check first to see if one function is the differential of the other (apart from a chain rule adjustment).

This method also works for integrals such as $\int \sqrt{3x+1}\ dx$

Here the product is $\sqrt{3x+1}$ and 1

$$\int \sqrt{3x+1}\ dx \ = \ \int (3x+1)^{\frac{1}{2}}\ dx$$

Start with $(3x+1)^{\frac{3}{2}}$ and see what happens when you differentiate

$$\frac{d}{dx}\left[(3x+1)^{\frac{3}{2}}\right] \ = \ \frac{3}{2}(3x+1)^{\frac{1}{2}}(3)$$

$$= \ \frac{9}{2}(3x+1)^{\frac{1}{2}}$$

This result is the original integral apart from the $\frac{9}{2}$

Hence $\int (3x+1)^{\frac{1}{2}}\ dx = \frac{2}{9}(3x+1)^{\frac{3}{2}} + c$

Check : Let $y = \frac{2}{9}(3x+1)^{\frac{3}{2}}$

Then $\frac{dy}{dx} = \frac{2}{9}\left[\frac{3}{2}(3x+1)^{\frac{1}{2}}(3)\right]$

$= (3x+1)^{\frac{1}{2}}$

Example : Given $\frac{dy}{dx} = \tan^4 3x \sec^2 3x$

Find y if $y = 1$ when $x = \frac{\pi}{12}$

Step 1 : Find $\int \tan^4 3x \sec^2 3x\ dx$

Note : the differential of $\tan x$ is $\sec^2 x$

Let $y = (\tan 3x)^5$

Then $\frac{dy}{dx} = 5(\tan 3x)^4 (3\sec^2 3x)$

$= 15\tan^4 3x \sec^2 3x$

Hence $\int \tan^4 3x \sec^2 3x\ dx = \frac{1}{15}(\tan 3x)^5 + c$

Step 2 : Substitute $y = 1$ and $x = \dfrac{\pi}{12}$ to find c

$$y = \frac{1}{15}(\tan 3x)^5 + c$$

$$1 = \frac{1}{15}\left(\tan\frac{\pi}{4}\right)^5 + c$$

$$1 = \frac{1}{15}(1)^5 + c$$

$$c = \frac{14}{15}$$

$$y = \frac{1}{15}(\tan 3x)^5 + \frac{14}{15}$$

Example :

The acceleration of an object is given by $a = \dfrac{5}{(3t+1)^3}$ for $t \geq 0$

where t = time in seconds
a = acceleration in meters per second per second

If the object is initially stationary then find the velocity of the object after $\dfrac{4}{3}$ seconds.

Step 1 : Since $a = \dfrac{dv}{dt}$ where v = velocity

we need to find $\displaystyle\int \frac{5}{(3t+1)^3}\,dt$

$$v = \int 5(3t+1)^{-3}\,dt$$

Start with $v = (3t+1)^{-2}$ and see what happens when you find $\dfrac{dv}{dt}$

$$v = (3t+1)^{-2} \Rightarrow \frac{dv}{dt} = -2(3t+1)^{-3}(3)$$
$$= -6(3t+1)^{-3}$$

This is almost the answer that we need.

If we change $v = (3t+1)^{-2}$ into $v = -\dfrac{5}{6}(3t+1)^{-2}$

then we get the required answer.

Hence $\displaystyle\int 5(3t+1)^{-3}\,dt = -\frac{5}{6}(3t+1)^{-2} + c$

$$v = -\frac{5}{6}(3t+1)^{-2} + c$$

Step 2 : Find c

The word *"initially"* implies $t = 0$

Substitute $v = 0$ *when* $t = 0$

$$0 = -\frac{5}{6}(0+1)^{-2} + c$$

$$c = \frac{5}{6}$$

Hence $v = -\frac{5}{6}(3t+1)^{-2} + \frac{5}{6}$

Step 3 : Find v when $t = \frac{4}{3}$

$$v = -\frac{5}{6}(3t+1)^{-2} + \frac{5}{6}$$

$$= -\frac{5}{6}(4+1)^{-2} + \frac{5}{6}$$

$$= -\frac{5}{6}\left(\frac{1}{25}\right) + \frac{5}{6}$$

$$= 0.8 \text{ m/s}$$

If you differentiate $y = x^2$ you are finding the gradient of the curve.
If you integrate $y = x^2$ you are finding the area under the curve.

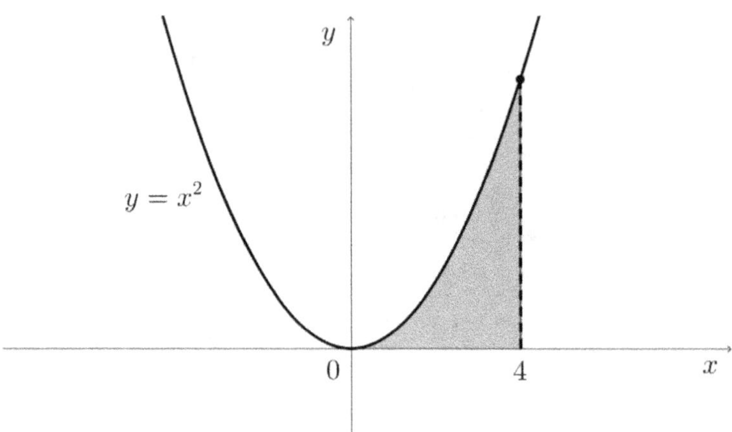

The shaded area is given by $\int_0^4 x^2 \, dx$

This is calculated in the following manner :

$$\int_0^4 x^2 \, dx = \left[\frac{x^3}{3}\right]_0^4$$

$$= \left[\frac{4^3}{3}\right] - \left[\frac{0^3}{3}\right]$$

$$= \frac{64}{3}$$

There is no need to include c when integrating here.

This process is called “integrating with limits”.

Example :

Find $\int_0^{\frac{\pi}{2}} \sin 2x \, dx$ and describe what you are finding.

Step 1 : $\int_0^{\frac{\pi}{2}} \sin 2x \, dx = \left[-\frac{1}{2}\cos 2x\right]_0^{\frac{\pi}{2}}$

$$= \left[-\frac{1}{2}\cos \pi\right] - \left[-\frac{1}{2}\cos 0\right]$$

$$= \left[-\frac{1}{2}(-1)\right] - \left[-\frac{1}{2}(1)\right]$$

$$= \left[\frac{1}{2}\right] - \left[-\frac{1}{2}\right]$$

$$= 1$$

Step 2 : Draw a diagram showing $y = \sin 2x$

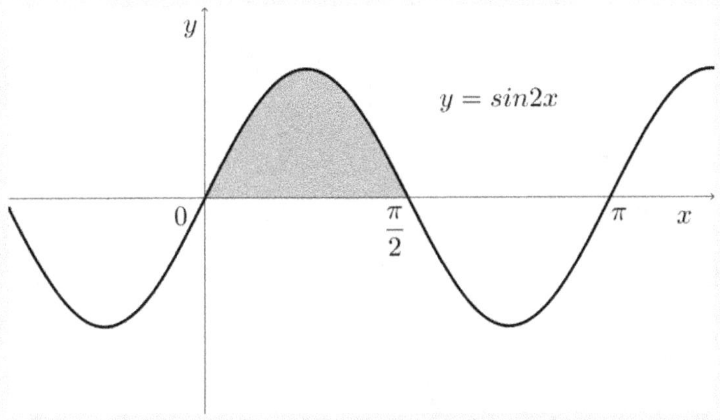

You are finding the area of the shaded part of the graph.

When you are asked to find an area using integration it is important that you draw a diagram showing the area required. This is because any area below the x-axis will give a negative result when you integrate the function.

Consider the graph of $y = x^3$

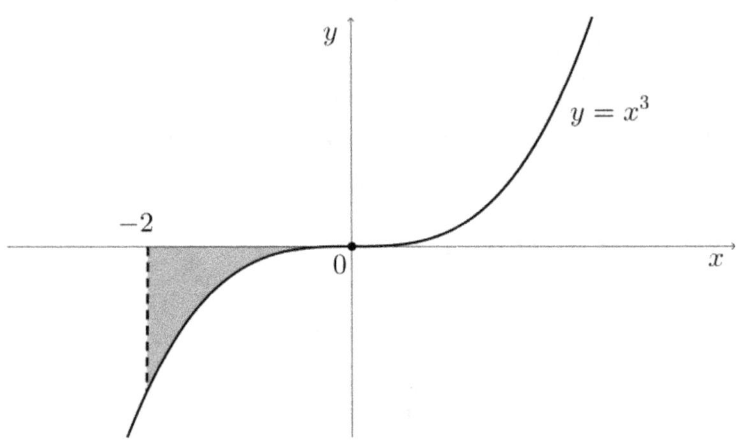

$$\int_{-2}^{0} x^3 \, dx = \left[\frac{x^4}{4}\right]_{-2}^{0}$$

$$= \left[\frac{0}{4}\right] - \left[\frac{(-2)^4}{4}\right]$$

$$= -4$$

The area is 4 square units.
However the value of the integral is -4.

This becomes important when the area you need to find is both above and below the x-axis.

For example, consider the shaded area below :

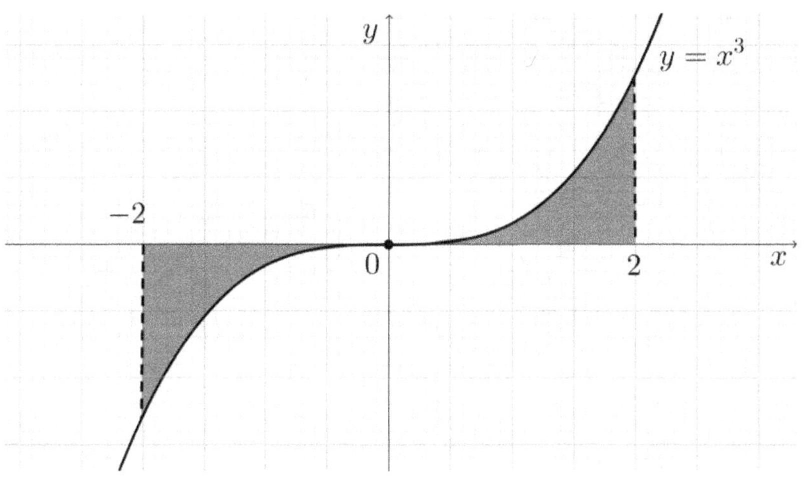

$$\int_{-2}^{2} x^3 \, dx = \left[\frac{x^4}{4}\right]_{-2}^{2}$$

$$= \left[\frac{(2)^4}{4}\right] - \left[\frac{(-2)^4}{4}\right]$$

$$= [4] - [4]$$

$$= 0$$

This is because the area above the x-axis is 4 and the area below is calculated by the integral as -4 .

The required answer is , of course , 8 square units.

Consequently you need to calculate areas above and below the x-axis separately.

Example :

The graph of the parabola $y = x^2 - 4x$ is shown below.
Calculate the shaded area.

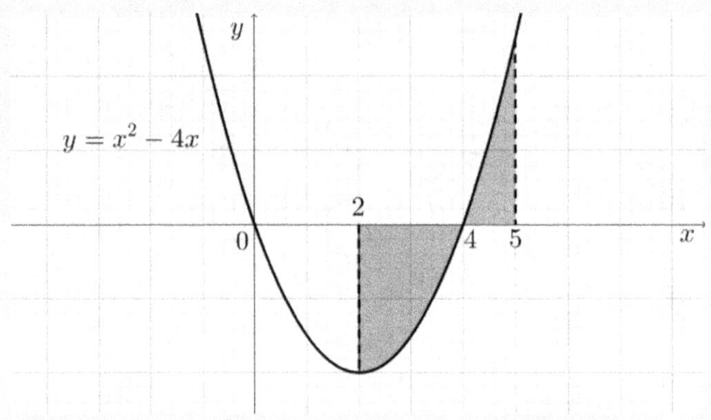

Step 1 : Find the shaded area that is below the x-axis

$$\int_2^4 x^2 - 4x \; dx = \left[\frac{x^3}{3} - 2x^2\right]_2^4$$

$$= \left[\frac{4^3}{3} - 2(4)^2\right] - \left[\frac{2^3}{3} - 2(2)^2\right]$$

$$= \left[-\frac{32}{3}\right] - \left[-\frac{16}{3}\right]$$

$$= -\frac{16}{3}$$

Hence the area is $\dfrac{16}{3}$ square units.

Step 2 : Find the shaded area that is above the x-axis

$$\int_4^5 x^2 - 4x \ dx = \left[\frac{x^3}{3} - 2x^2\right]_4^5$$

$$= \left[\frac{5^3}{3} - 2(5)^2\right] - \left[\frac{4^3}{3} - 2(4)^2\right]$$

$$= \left[-\frac{25}{3}\right] - \left[-\frac{32}{3}\right]$$

$$= \frac{7}{3}$$

Hence the area is $\frac{7}{3}$ square units.

Step 3 : Add the two areas

$$\frac{16}{3} + \frac{7}{3} = \frac{23}{3}$$

Hence the total shaded area is $\frac{23}{3}$ square units

Note : $\int_2^5 x^2 - 4x \ dx = -3$

This is the sum of $-\frac{16}{3}$ and $\frac{7}{3}$

Exercise 5

1) Integrate the following functions and check your answer by differentiation.

(a) $6x^2+8x+3$ (b) $\frac{9}{x^2}$ (c) $\frac{6}{\sqrt{x^3}}$

(d) $12\sqrt[3]{x}$ (e) $\frac{3}{4x^5}$ (f) $\frac{18x^6+1}{x}$

2) Find the following integrals and check your answer by differentiation.

(a) $\int \frac{x}{x^2+1}\,dx$ (b) $\int \cos 4x\,dx$ (c) $\int \sec^2 3x\,dx$

(d) $\int e^{5x}\,dx$ (e) $\int \sin^3 x \cos x\,dx$ (f) $\int xe^{5x^2}\,dx$

(g) $\int (3x+2)^5\,dx$ (h) $\int \sqrt{2x-1}\,dx$ (i) $\int \frac{1}{\sqrt{2x-1}}\,dx$

3) (a) Find $\int \frac{6x+4}{3x^2+4x+5}\,dx$

(b) Find $\int \frac{\cos x}{\sin x}\,dx$ and hence write down the integral of $\cot x$.

4) Find the exact value of the following definite integrals.

(a) $\int_2^8 \frac{1}{x+1}\,dx$ (b) $\int_0^{\frac{\pi}{2}} \sin 2x\,dx$ (c) $\int_{\frac{\pi}{6}}^{\frac{\pi}{4}} \left(-2\operatorname{cosec}2x\cot 2x\right)dx$

(d) $\int_0^1 e^{2x}\,dx$ (e) $\int_{-1}^1 3x^2+2x+1\,dx$ (f) $\int_{\frac{\pi}{4}}^{\frac{\pi}{3}} \tan^3 x\sec^2 x\,dx$

5) (a) $\int_{-2}^{2} 4x^3 \, dx$

(b) Find the area enclosed by the curve $y = 4x^3$, the lines $x = -2$ and $x = 2$, and the x-axis.

6) Find $\int_{0}^{2\pi} \sin x \, dx$ and comment on your answer by using

the graph of $y = \sin x$

7) The acceleration of an object is given by $a = 80t^3$ for $t \geq 0$ where $t =$ time in seconds $a =$ acceleration in meters per second per second.

If the object is initially stationary at the origin then find

(a) the velocity of the object after 2 seconds.

(b) the distance moved during the first 2 seconds.

8) In the following diagram the shaded area is 3 square units.

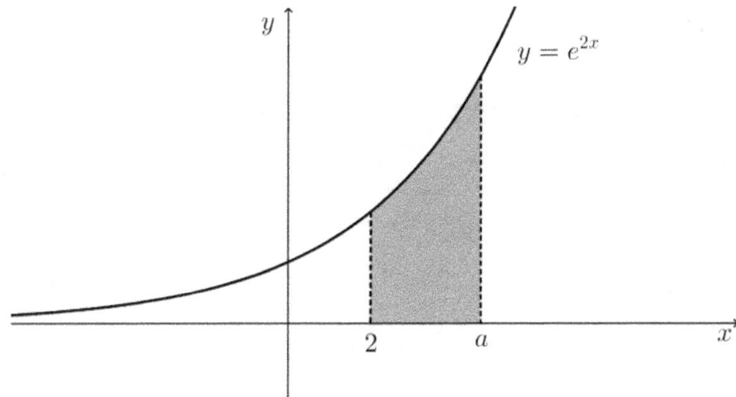

Find the exact value of a

9) By letting $\tan x = \dfrac{\sin x}{\cos x}$ find $\int \tan x \, dx$

10) The velocity of an object is given by $v = 12t - 3t^2$ for $t \geq 0$ where $t =$ time in seconds and v is measured in meters per second. The object is initially stationary at the origin.

(a) Sketch a graph of $v = 12t - 3t^2$ for $t \geq 0$

Note : the area under a velocity-time graph represents distance.

(b) the distance moved during the first 4 seconds.

(c) the distance moved during the first 6 seconds.

(d) the displacement of the object during the first 6 seconds.

(e) write down an integral that represents the distance moved during the first 6 seconds.

(f) write down an integral that represents the displacement during the first 6 seconds.

11) Given that $y = f(x)$ has gradient function $\dfrac{dy}{dx} = 8x + 3$

Find y if $f(1) = 9$

12) Given that $y = f(x)$ has gradient function $\dfrac{dy}{dx} = \dfrac{2x}{x^2 + 1}$

Find y if $f(0) = 2$

13) (a) $\displaystyle\int_0^4 \frac{3}{3x+2} \, dx = \ln k$. Find the value of k

(b) Show that $\displaystyle\int \frac{1}{2x+1} \, dx = \ln A\sqrt{2x+1}$

where A is a constant.

14) (a) Differentiate $y = x\ln x - x$

(b) Hence write down $\int \ln x \; dx$

15) Using the information given on the graph below

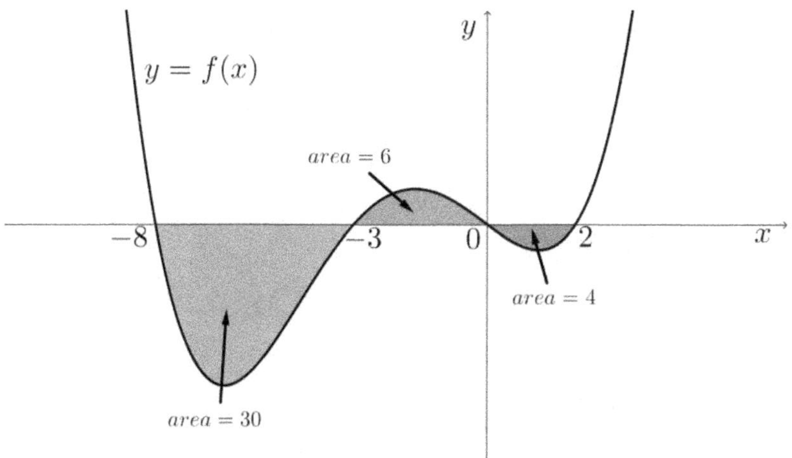

Calculate (a) $\int_{-8}^{-3} f(x)\; dx$ (b) $\int_{-8}^{2} f(x)\; dx$ (c) $\int_{-3}^{2} f(x)\; dx$

16) (a) Write $\sin^2 x$ in terms of $\cos 2x$

(b) Hence find $\int_{0}^{\frac{\pi}{6}} \sin^2 x \; dx$

17) (a) Show that $\sin^5 x = \sin x - 2\cos^2 x \sin x + \cos^4 x \sin x$

(b) Hence find $\int \sin^5 x \; dx$

18) (a) $\int \tan^2 x \; dx$ (Hint: use the Pythagorean trig identities)

(b) $\int \cos^3 x \; dx$ (Hint: use the identity $\cos^2 x + \sin^2 x = 1$)

19) (a) Find $\int \frac{\ln x}{x} \; dx$

(b) Hence find the exact value of $\int_1^5 \frac{\ln x}{x} \; dx$

(c) Use a GDC to draw a sketch of $y = \frac{\ln x}{x}$

Shade the area found in part (b) and check your answer by using the CALC function.

20) For $y = f(x)$ you are given that $\frac{d^2 y}{dx^2} = -\frac{3}{x^2}$

and the gradient of $f(x)$ at the point (1, 5) is 4.

Find $f(x)$

Integration by Substitution

It is not always possible to integrate using observation and "reverse differentiation".

For example $\int \frac{x+1}{x-1}\, dx$ does not have any obvious solution.

The technique "integration by substitution" is a method used to convert such integrals into a simpler form that can be easily integrated.

Here we let $u = x-1$ (the denominator of the integral) and then change everything into terms containing u.

If $u = x-1$ then $x = u+1$ and $\frac{dx}{du} = 1$

This means we can write $dx = 1\, du$ or $dx = du$

Hence
$$\int \frac{x+1}{x-1}\, dx = \int \frac{(u+1)+1}{u}\, du$$
$$= \int \frac{u+2}{u}\, du$$
$$= \int 1+\frac{2}{u}\, du$$

This can be easily integrated by using the previous method of "reverse differentiation".

$$\int 1+\frac{2}{u}\, du = u + 2\ln u + c$$

The last step is to change everything back into x and then simplify

$$u + 2\ln u + c = (x-1) + 2\ln(x-1) + c$$

$$= x + \ln(x-1)^2 + c - 1$$

$$= x + \ln(x-1)^2 + k$$

Note : Since c is a constant , $c-1$ is also a constant.
This can be simplified to the constant k.

You could also write $x + \ln(x-1)^2 + k$ as $x + \ln(x-1)^2 + \ln A$ where A is another constant term.

This gives the neater answer of $x + \ln A(x-1)^2$

Example : Find $\int \frac{x}{3x+1}\, dx$

Step 1 : Let $u = 3x+1$

Then $x = \frac{1}{3}(u-1)$ and $\frac{dx}{du} = \frac{1}{3}$

Hence $dx = \frac{1}{3}\, du$

Step 2 : Replace all x terms with u terms

$$\int \frac{x}{3x+1}\, dx = \int \frac{\frac{1}{3}(u-1)}{u} \frac{1}{3}\, du$$

$$= \frac{1}{9} \int \frac{u-1}{u}\, du$$

$$= \frac{1}{9}\int 1 - \frac{1}{u} \; du$$

$$= \frac{1}{9}\left[u - \ln u\right] + c$$

Step 3 : Convert back to x terms and simplify

$$\frac{1}{9}\left[u - \ln u\right] + c = \frac{1}{9}\left[(3x+1) - \ln(3x+1)\right] + c$$

$$= \frac{1}{3}x + \frac{1}{9} - \frac{1}{9}\ln(3x+1) + c$$

$$= \frac{1}{3}x - \frac{1}{9}\ln(3x+1) + k$$

Note : $\int k\,f(x)\;dx = k\int f(x)\;dx$ where k is a constant

If the previous example had included limits then there would be no need for Step 3 (converting back to x terms). Instead just convert the limits to u.

For example $\int_1^2 \frac{x}{3x+1}\;dx = \int_4^7 \frac{\frac{1}{3}(u-1)}{u}\,\frac{1}{3}\;du$

since $x = 1 \Rightarrow u = 4$ and $x = 2 \Rightarrow u = 7$

$$= \left[\frac{1}{9}\left[u - \ln u\right]\right]_4^7$$

$$= \left[\frac{1}{9}\left[7 - \ln 7\right]\right] - \left[\frac{1}{9}\left[4 - \ln 4\right]\right]$$

$$= \frac{1}{9}\left[3 + \ln\frac{4}{7}\right]$$

Example : Find $\int (x+3)(x-2)^8 \, dx$

Step 1 : Let $u = x - 2$

Then $x = u + 2$ and $\frac{dx}{du} = 1$

Hence $dx = du$

Step 2 : Replace all x terms with u terms

$$\int (x+3)(x-2)^8 \, dx = \int (u+5)(u)^8 \, du$$

$$= \int u^9 + 5u^8 \, du$$

$$= \frac{u^{10}}{10} + \frac{5u^9}{9} + c$$

Step 3 : Convert back to x terms and simplify

$$\frac{u^{10}}{10} + \frac{5u^9}{9} + c = \frac{(x-2)^{10}}{10} + \frac{5(x-2)^9}{9} + c$$

$$= \frac{(x-2)^9}{90}\left[9(x-2)+50\right] + c$$

$$= \frac{(x-2)^9}{90}[9x+32] + c$$

$$= \frac{(x-2)^9(9x+32)}{90} + c$$

Example : Find $\int \frac{x+1}{\sqrt{2x+1}}\ dx$

Step 1 : Let $u = 2x+1$

Then $x = \frac{1}{2}(u-1)$ and $\frac{dx}{du} = \frac{1}{2}$

Hence $dx = \frac{1}{2}du$

Step 2 : Replace all x terms with u terms

$$\int \frac{x+1}{\sqrt{2x+1}}\ dx = \int \frac{\frac{1}{2}(u-1)+1}{\sqrt{u}}\ \frac{1}{2}du$$

$$= \frac{1}{2}\int \frac{\frac{1}{2}u - \frac{1}{2} + 1}{\sqrt{u}}\ du$$

$$= \frac{1}{2}\int \frac{\frac{1}{2}u + \frac{1}{2}}{\sqrt{u}}\ du$$

$$= \frac{1}{4}\int \frac{u+1}{\sqrt{u}}\ du$$

$$= \frac{1}{4}\int u^{\frac{1}{2}} + u^{-\frac{1}{2}}\ du$$

$$= \frac{1}{4}\left[\frac{u^{\frac{3}{2}}}{\frac{3}{2}} + \frac{u^{\frac{1}{2}}}{\frac{1}{2}}\right] + c$$

$$= \frac{1}{4}\left[\frac{2}{3}u^{\frac{3}{2}} + 2u^{\frac{1}{2}}\right] + c$$

$$= \frac{1}{6}\left[u^{\frac{3}{2}} + 3u^{\frac{1}{2}}\right] + c$$

$$= \frac{1}{6}\,u^{\frac{1}{2}}[u + 3] + c$$

Step 3 : Convert back to x terms and simplify

$$\frac{1}{6}\,u^{\frac{1}{2}}[u + 3] + c = \frac{1}{6}\,(2x+1)^{\frac{1}{2}}\left[\,(2x+1) + 3\right] + c$$

$$= \frac{1}{6}\,(2x+1)^{\frac{1}{2}}[\,2x+4] + c$$

$$= \frac{1}{3}\,(2x+1)^{\frac{1}{2}}(x+2) + c$$

A slight variation of this method can be used to find integrals like

$$\int \frac{1}{\sqrt{1-x^2}}\,dx \quad \text{and} \quad \int \frac{1}{1+x^2}\,dx$$

Here we substitute using trigonometric functions.

For $\int \frac{1}{\sqrt{1-x^2}}\,dx$ we use $x = \sin u$

This gives $\frac{dx}{du} = \cos u$ and hence $dx = \cos u \; du$

The integral then becomes $\int \frac{1}{\sqrt{1-\sin^2 x}} \cos u \; du$

Since $1-\sin^2 x = \cos^2 x$ everything simplifies down to $\int 1 \; du$ giving an answer of $u + c$. This converts into $\arcsin x + c$

$\int \frac{1}{\sqrt{1-x^2}}\,dx = \arcsin x + c$ is a standard integral

that can be found in the IB formula booklet.

You are allowed to quote the answer without proof.

Example : Find $\int \frac{1}{1+x^2}\ dx$

Step 1 : Let $x = \tan u$

Then $\frac{dx}{du} = \sec^2 u$

Hence $dx = \sec^2 u\ \ du$

Step 2 : Replace all x terms with u terms

$$\int \frac{1}{1+x^2}\ dx = \int \frac{1}{1+\tan^2 u}\ \sec^2 u\ \ du$$

$$= \int \frac{1}{\sec^2 u}\ \sec^2 u\ \ du$$

$$= \int 1\ du$$

$$= u + c$$

$$= \arctan x + c$$

$\int \frac{1}{1+x^2}\ dx = \arctan x + c$ is also a standard integral

Example : Find $\int \frac{1}{\sqrt{9-25x^2}}\,dx$

Step 1 : Let $x = \frac{3}{5}\sin u$

Then $\frac{dx}{du} = \frac{3}{5}\cos u$

Hence $dx = \frac{3}{5}\cos u \; du$

Step 2 : Replace all x terms with u terms

$$\int \frac{1}{\sqrt{9-25x^2}}\,dx = \int \frac{1}{\sqrt{9-25\left(\frac{3}{5}\sin u\right)^2}}\;\frac{3}{5}\cos u \;\; du$$

$$= \frac{3}{5}\int \frac{1}{\sqrt{9-25\left(\frac{9}{25}\sin^2 u\right)}}\;\cos u \;\; du$$

$$= \frac{3}{5}\int \frac{1}{\sqrt{9-9\sin^2 u}}\;\cos u \;\; du$$

$$= \frac{3}{5}\int \frac{1}{\sqrt{9(1-\sin^2 u)}}\;\cos u \;\; du$$

$$= \frac{3}{5}\int \frac{1}{\sqrt{9\cos^2 u}}\;\cos u \;\; du$$

$$= \frac{3}{5}\int \frac{1}{3\cos u} \cos u \quad du$$

$$= \frac{1}{5}\int 1 \; du$$

$$= \frac{1}{5}u + c$$

$$= \frac{1}{5}\arcsin\frac{5}{3}x + c$$

Since $\arcsin x$ has domain $-1 \le x \le 1$ the domain here is

$$-1 < \frac{5}{3}x < 1 \quad \text{or } -\frac{3}{5} < x < \frac{3}{5}$$

Note : the inequality changed from $\le$ to $<$ because

$$\int \frac{1}{\sqrt{9-25x^2}} \; dx \text{ doesn't exist if } x = \pm\frac{3}{5}$$

Also the domain of $\dfrac{1}{\sqrt{9-25x^2}}$ is $9-25x^2 > 0$

or $-\frac{3}{5} < x < \frac{3}{5}$

The Standard Integrals

$$\int \frac{1}{\sqrt{a^2 - x^2}}\,dx = \arcsin\left(\frac{x}{a}\right) + c$$

$$\int \frac{1}{a^2 + x^2}\,dx = \frac{1}{a}\arctan\left(\frac{x}{a}\right) + c$$

The previous question could have been answered by using the Standard Integral (after performing some algebraic manipulation)

$$\int \frac{1}{\sqrt{9 - 25x^2}}\,dx = \int \frac{1}{\sqrt{25\left(\frac{9}{25} - x^2\right)}}\,dx$$

$$= \frac{1}{5}\int \frac{1}{\sqrt{\frac{9}{25} - x^2}}\,dx$$

$$= \frac{1}{5}\int \frac{1}{\sqrt{\left(\frac{3}{5}\right)^2 - x^2}}\,dx$$

This is now in the correct format of $\int \frac{1}{\sqrt{a^2 - x^2}}\,dx$ where $a = \frac{3}{5}$

Hence $\frac{1}{5}\int \frac{1}{\sqrt{\left(\frac{3}{5}\right)^2 - x^2}}\,dx = \frac{1}{5}\arcsin\frac{5}{3}x + c$

Example : Use the Standard Integral $\int \frac{1}{a^2+x^2}\,dx = \frac{1}{a}\arctan\left(\frac{x}{a}\right) + c$

to find $\int \frac{1}{x^2+4x+20}\,dx$

Step 1 : Write $x^2+4x+20$ in the form $a(x-h)^2+k$ by "completing the square".

$$\begin{aligned} x^2+4x+20 &= x^2+4x+4+16 \\ &= (x+2)^2+16 \end{aligned}$$

Step 2 : $$\begin{aligned} \int \frac{1}{x^2+4x+20}\,dx &= \int \frac{1}{16+(x+2)^2}\,dx \\ &= \int \frac{1}{4^2+(x+2)^2}\,dx \\ &= \frac{1}{4}\arctan\left(\frac{x+2}{4}\right) + c \end{aligned}$$

A little more work is required if the x^2 coefficient is not 1

For example $\displaystyle\int \frac{1}{9x^2+6x+26}\,dx = \int \frac{1}{(3x+1)^2+25}\,dx$

$$= \int \frac{1}{5^2+(3x+1)^2}\,dx$$

Now you need to let $u = 3x+1$ to get the required format.

This means $\dfrac{du}{dx} = 3$ and $dx = \dfrac{1}{3}du$

Then $\displaystyle\int \frac{1}{5^2+(3x+1)^2}\,dx = \int \frac{1}{5^2+(u)^2}\,\frac{1}{3}\,du$

$$= \frac{1}{3}\int \frac{1}{5^2+u^2}\,du$$

$$= \frac{1}{3}\left[\frac{1}{5}\arctan\left(\frac{u}{5}\right)\right] + c$$

$$= \frac{1}{15}\arctan\left(\frac{3x+1}{5}\right) + c$$

Integration by Parts

This is most commonly used for products of different types of functions.

For example $\int x\cos x \; dx$ and $\int x^2 e^x \; dx$

Derivation of the 'integration by parts' formula :

Using the product rule for differentiation we have

$$\frac{d}{dx}(uv) = u\frac{dv}{dx} + v\frac{du}{dx}$$

Integrating both sides gives

$$uv = \int \left[u\frac{dv}{dx} + v\frac{du}{dx} \right] \; dx$$

$$uv = \int u\frac{dv}{dx} \; dx \; + \; \int v\frac{du}{dx} \; dx$$

Rearranging $\int u\frac{dv}{dx} \; dx = uv - \int v\frac{du}{dx} \; dx$

When using this formula it is important to choose the u and the $\frac{dv}{dx}$ correctly.

For example when finding $\int x\cos x \; dx$ you must let $u = x$ and $\frac{dv}{dx} = \cos x$

This will result in $\frac{du}{dx} = 1$ and $v = \sin x$

$$\int u\frac{dv}{dx} \; dx = uv - \int v\frac{du}{dx} \; dx$$

$$\int x\cos x \; dx = x\sin x - \int (\sin x)(1) \; dx$$

$$= x\sin x + \cos x + c$$

Sometimes you will need to apply the integration by parts formula more than once. For example

$$\int x^2 e^x \; dx = x^2 e^x - \int e^x (2x) \; dx$$

$$= x^2 e^x - 2\int xe^x \; dx$$

$$= x^2 e^x - 2\left[xe^x - \int e^x (1) \; dx\right]$$

$$= x^2 e^x - 2xe^x + 2\int e^x \; dx$$

$$= x^2 e^x - 2xe^x + 2e^x + c$$

Example : Find $\int x\sec^2 x \, dx$

Step 1 : Let $u = x$ and $\dfrac{dv}{dx} = \sec^2 x$

Then $\dfrac{du}{dx} = 1$ and $v = \tan x$

Step 2 : Substitute into the formula

$$\int u\frac{dv}{dx} \, dx = uv - \int v\frac{du}{dx} \, dx$$

$$\int x\sec^2 x \, dx = x\tan x - \int (\tan x)(1) \, dx$$

$$= x\tan x - \int \frac{\sin x}{\cos x} \, dx$$

$$= x\tan x + \ln(\cos x) + c$$

The integration by parts formula is also used to find integrals such as

$$\int \ln x \, dx \quad \text{and} \quad \int \arcsin x \, dx$$

In these cases the u is always the given function and $\frac{dv}{dx} = 1$

For example :

$$\int \ln x \, dx = \int (\ln x)(1) \, dx$$

$$= (\ln x)(x) - \int (x)\left(\frac{1}{x}\right) dx$$

$$= x \ln x - \int 1 \, dx$$

$$= x \ln x - x + c$$

Example : Find $\int \arcsin x \, dx$

Step 1 : Let $u = \arcsin x$ and $\frac{dv}{dx} = 1$

Then $\frac{du}{dx} = \frac{1}{\sqrt{1-x^2}}$ and $v = x$

Step 2 : Substitute into the formula

$$\int u \frac{dv}{dx} \, dx = uv - \int v \frac{du}{dx} \, dx$$

$$\int (\arcsin x)(1) \, dx = (\arcsin x)(x) - \int (x)\left(\frac{1}{\sqrt{1-x^2}}\right) dx$$

$$= x \arcsin x - \int x\left(1-x^2\right)^{-\frac{1}{2}} dx$$

$$= x \arcsin x + \left(1-x^2\right)^{\frac{1}{2}} + c$$

$$= x \arcsin x + \sqrt{1-x^2} + c$$

Another variation on *integration by parts* occurs when you integrate functions such as $e^x \sin x$

$$\int e^x \sin x \; dx \;=\; e^x(-\cos x) - \int (-\cos x)(e^x) \; dx$$

$$= -\,e^x \cos x + \int e^x \cos x \, dx$$

$$= \; -\,e^x \cos x + \left[e^x \sin x - \int (\sin x)(e^x) \, dx \right]$$

$$= -\,e^x \cos x + e^x \sin x - \int e^x \sin x \, dx$$

At this point any further integration will end up going around in circles.
The method here is to rearrange the terms

$$\int e^x \sin x \; dx \;=\; -\,e^x \cos x + e^x \sin x - \int e^x \sin x \, dx$$

Add $\int e^x \sin x \; dx$ to both sides of the equation

$$2\int e^x \sin x \; dx \;=\; -\,e^x \cos x + e^x \sin x$$

$$\int e^x \sin x \; dx \;=\; \frac{1}{2}\left(-\,e^x \cos x + e^x \sin x\right)$$

Hence $\int e^x \sin x \; dx \;=\; \frac{1}{2} e^x (\sin x - \cos x) \; + \; c$

Integration by using Partial Fractions

When you are asked to find an integral like $\int \frac{2x+1}{x^2+x-2}\ dx$

the first step is always to see if the top is the differential of the bottom

(or some multiple of the differential of the bottom).

Here you should notice that $\frac{d}{dx}\left(x^2+x-2\right) = 2x+1$

Hence $\int \frac{2x+1}{x^2+x-2}\ dx = \ln\left(x^2+x-2\right)+c$

Note : technically you should write $\ln\left|x^2+x-2\right|+c$
but this is never penalized in the exam.

Now what happens if the top is NOT a multiple of the differential of the bottom.

For example $\int \frac{2x+3}{x^2+x-2}\ dx$

Here you use **partial fractions.**

The first step is to factorize the denominator :

$$\frac{2x+3}{x^2+x-2} = \frac{2x+3}{(x+2)(x-1)}$$

Then let $\dfrac{2x+3}{(x+2)(x-1)} = \dfrac{A}{(x+2)}+\dfrac{B}{(x-1)}$

Next find the values of A and B

Multiply both sides by $(x+2)(x-1)$ to give

$$2x+3 = A(x-1)+B(x+2)$$

When $x=1$, $\quad 5=3B \quad \Rightarrow \quad B=\dfrac{5}{3}$

When $x=-2$, $\quad -1=-3A \quad \Rightarrow \quad A=\dfrac{1}{3}$

Hence $\dfrac{2x+3}{x^2+x-2} = \dfrac{1}{3(x+2)}+\dfrac{5}{3(x-1)}$ and

$$\int \frac{2x+3}{x^2+x-2}\,dx = \int \frac{1}{3(x+2)}+\frac{5}{3(x-1)}\,dx$$

$$= \frac{1}{3}\int \frac{1}{(x+2)}+\frac{5}{(x-1)}\,dx$$

$$= \frac{1}{3}\left[\ln(x+2)+5\ln(x-1)\right] + c$$

$$= \frac{1}{3}\left[\ln(x+2)(x-1)^5\right] + c$$

$$= \ln\sqrt[3]{(x+2)(x-1)^5} + c$$

$$= \ln A\sqrt[3]{(x+2)(x-1)^5}$$

Exercise 6

1) Integrate the following functions by using a suitable substitution

(a) $\dfrac{x+2}{x-2}$ (b) $(2x+3)(2x+1)^5$ (c) $\dfrac{x}{\sqrt{3x-1}}$

2) Find the following integrals using integration by parts

(a) $\int x\sin 2x \; dx$ (b) $\int xe^{3x} \; dx$ (c) $\int x\ln x \, dx$

(d) $\int x^2 \cos x \; dx$ (e) $\int x^2 e^{4x} \; dx$

3) (a) Write $\dfrac{14}{x^2+3x-10}$ in partial fractions

(b) Hence find $\int \dfrac{14}{x^2+3x-10} \; dx$

4) Find $\int \dfrac{1}{\sqrt{16-x^2}} \; dx$ by letting $x = 4\sin u$

5) Find $\int \dfrac{1}{4+9x^2} \; dx$ by letting $x = \dfrac{2}{3}\tan u$

6) Use the standard integral $\int \dfrac{1}{\sqrt{a^2-x^2}} \; dx = \arcsin\left(\dfrac{x}{a}\right) + c$

to find $\int \dfrac{1}{\sqrt{1-9x^2}} \; dx$

7) Use the standard integral $\int \frac{1}{a^2+x^2}\,dx = \frac{1}{a}\arctan\left(\frac{x}{a}\right) + c$

to find $\int \frac{1}{25+49x^2}\,dx$

8) (a) Differentiate $y = \arctan x$

(b) Use your answer to part (a) to help find $\int \arctan x\,dx$

9) $\int \arccos 3x\,dx$

10) Find the exact value of the following definite integrals.

(a) $\int_0^{\frac{\pi}{3}} x\cos x\,dx$ (b) $\int_1^2 \frac{x+1}{2x-1}\,dx$

11) (a) $\int e^x \sin x\,dx$

(b) Hence find $\int_0^{\pi} e^x \sin x\,dx$

12) (a) Write $3-2x-x^2$ in the form $a(x-h)^2+k$

(b) Hence find $\int \frac{1}{\sqrt{3-2x-x^2}}\,dx$

13) (a) Write $4x^2 - 4x + 17$ in the form $a(x-h)^2 + k$

(b) Hence find $\int \frac{1}{4x^2 - 4x + 17}\, dx$

14) (a) (i) Differentiate $x^2 - 4x + 3$

(ii) Hence find $\int \frac{x-2}{x^2 - 4x + 3}\, dx$

(b) (i) Write $\frac{x-2}{x^2 - 4x + 3}$ in partial fractions.

(ii) Hence find $\int \frac{x-2}{x^2 - 4x + 3}\, dx$

(c) Show that your two answers are identical.

The area between two curves

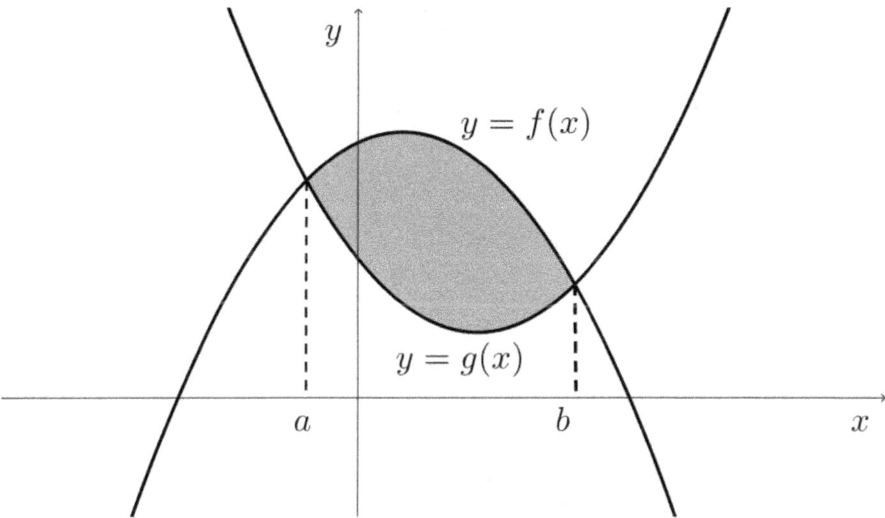

The shaded area could be calculated by doing two separate integrals

$$\int_a^b f(x)\,dx \;-\; \int_a^b g(x)\,dx$$

However it is often easier to combine the functions and calculate

$$\int_a^b f(x) - g(x)\;\;dx$$

For example , to calculate the area shaded below between

$$f(x) = -x^2 + 2x + 6 \qquad \text{and} \qquad g(x) = x^2 - 4x - 2$$

you should find $\int_a^b \left(-x^2 + 2x + 6\right) - \left(x^2 - 4x - 2\right) \; dx$

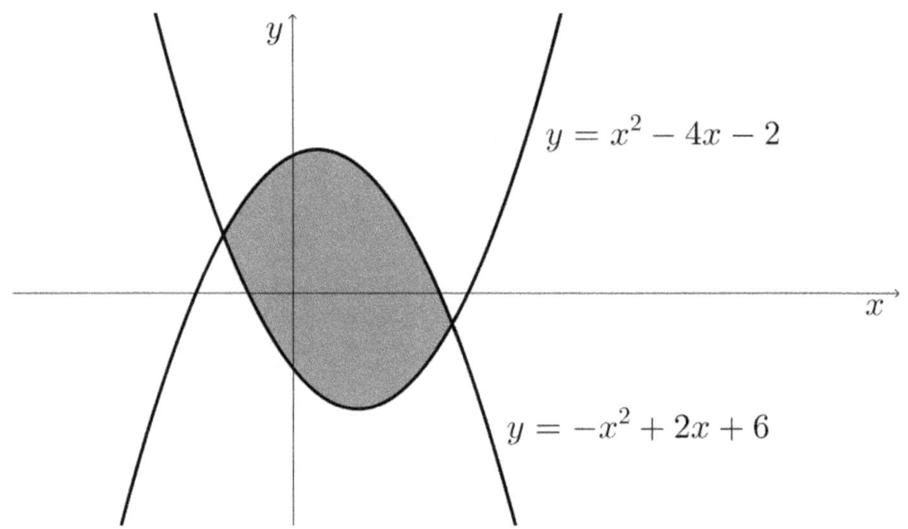

This simplifies down to

$$\int_a^b -2x^2 + 6x + 8 \quad dx \qquad \text{or} \qquad 2\int_a^b -x^2 + 3x + 4 \quad dx$$

Not only is this much less work but it also eliminates the need to deal with negative areas.

To find a and b you need to solve $f(x) = g(x)$

$$-x^2 + 2x + 6 = x^2 - 4x - 2$$

$$2x^2 - 6x - 8 = 0$$

$$x^2 - 3x - 4 = 0$$

$$(x+1)(x-4) = 0$$

$$x = -1 \ , \ 4$$

Then we have $2\int_{-1}^{4} -x^2 + 3x + 4 \quad dx = 2\left[-\frac{x^3}{3} + \frac{3x^2}{2} + 4x\right]_{-1}^{4}$

$$= 2\left[-\frac{4^3}{3} + \frac{3(4^2)}{2} + 4(4)\right] - 2\left[-\frac{(-1)^3}{3} + \frac{3(-1)^2}{2} + 4(-1)\right]$$

$$= 2\left[\frac{56}{3}\right] - 2\left[-\frac{13}{6}\right]$$

$$= \frac{125}{3}$$

Volume of Rotation

1) Rotation about the x-axis

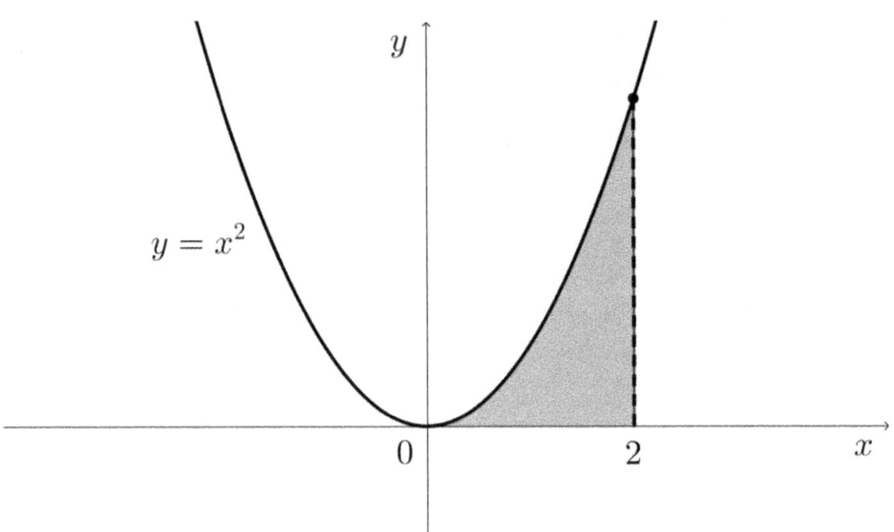

The volume formed when the shaded area is rotated 360° about the x-axis is given by the formula

$$\int_a^b \pi y^2 \, dx$$

where , in this case , $y = x^2$ and the limits a and b are 0 and 2

Hence $$\int_a^b \pi y^2 \, dx = \int_0^2 \pi \left(x^2\right)^2 \, dx$$

$$= \int_0^2 \pi x^4 \, dx$$

$$= \left[\frac{\pi x^5}{5}\right]_0^2$$

$$= \left[\frac{32\pi}{5}\right] - [0]$$

$$= \frac{32\pi}{5}$$

The shape of the resulting volume is shown below :

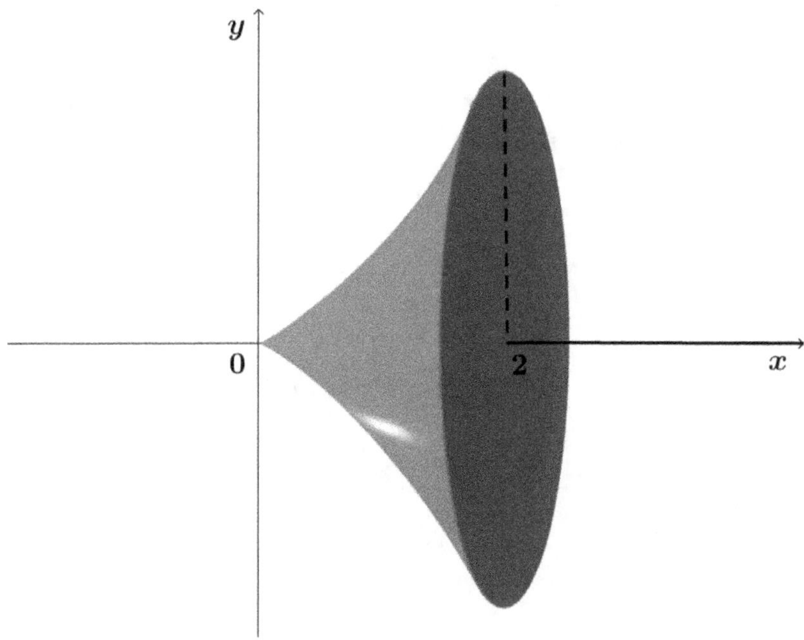

Example : The graph of the function $y = \sin x$ in the interval $[0, \pi]$ is rotated through 2π radians about the x-axis. Find the resulting volume.

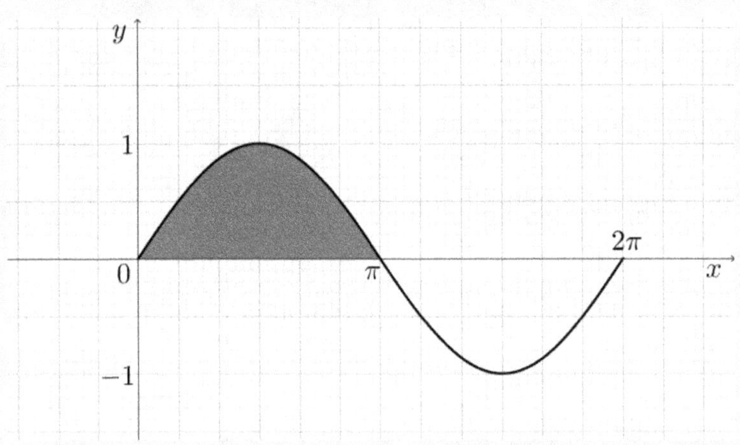

Using the formula $\int_a^b \pi y^2 \, dx$

Volume when shaded area is rotated about the x-axis

$$= \int_0^\pi \pi (\sin x)^2 \, dx$$

$$= \pi \int_0^\pi \sin^2 x \, dx$$

$$= \pi \int_0^\pi \frac{1}{2} (1 - \cos 2x) \, dx$$

$$= \frac{\pi}{2}\int_0^{\pi} 1-\cos 2x \;\; dx$$

$$= \frac{\pi}{2}\left[x-\frac{1}{2}\sin 2x\right]_0^{\pi}$$

$$= \frac{\pi}{2}\left[\pi-\frac{1}{2}\sin 2\pi\right] - \frac{\pi}{2}\left[0-\frac{1}{2}\sin 0\right]$$

$$= \frac{\pi}{2}\left[\pi-0\right] - \left[0\right]$$

$$= \frac{\pi^2}{2}$$

2) Rotation about the *y*-axis

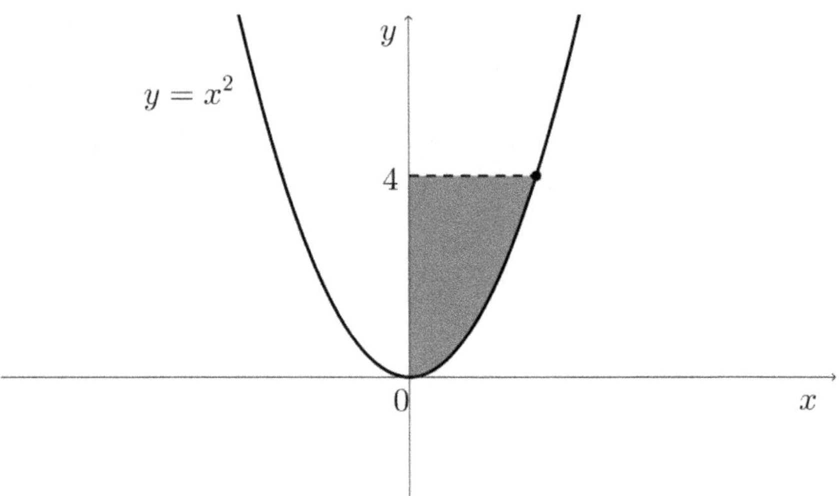

The volume formed when the shaded area is rotated 360° about the *y*-axis is given by the formula

$$\int_a^b \pi x^2 \, dy$$

where , in this case , $x^2 = y$ and the limits a and b are 0 and 4

Hence $$\int_a^b \pi x^2 \, dy = \int_0^4 \pi y \, dy$$

$$= \left[\frac{\pi y^2}{2} \right]_0^4$$

$$= \left[\frac{16\pi}{2}\right] - [0]$$

$$= 8\pi$$

The shape of the resulting volume is shown below :

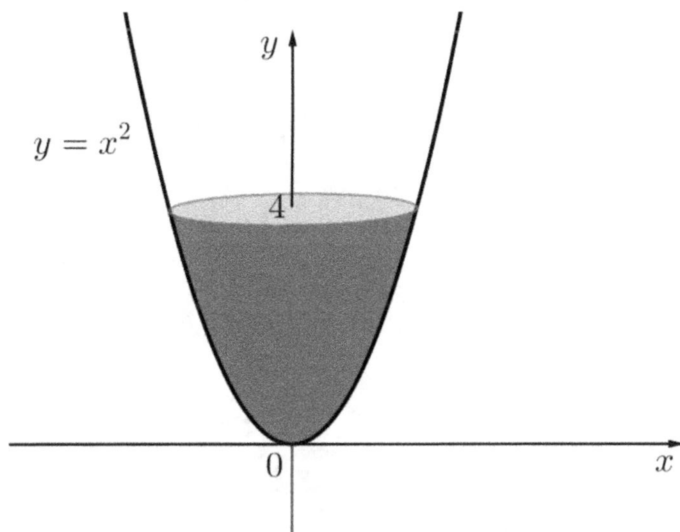

Example :

The graph of the function $y = \ln x$ is rotated through 2π radians about the y-axis from $y = 0$ to $y = 1$. Find the resulting volume.

Step 1 : Rewrite $y = \ln x$ as $x = e^y$

Step 2 : Volume $= \int_a^b \pi x^2 \, dy$

$$= \int_0^1 \pi \left(e^y\right)^2 \, dy$$

$$= \int_0^1 \pi e^{2y} \, dy$$

$$= \left[\frac{\pi e^{2y}}{2}\right]_0^1$$

$$= \left[\frac{\pi e^2}{2}\right] - \left[\frac{\pi}{2}\right]$$

$$= \frac{\pi}{2}\left[e^2 - 1\right]$$

Exercise 7

1) (a) Find the exact area enclosed by curve $y = \tan x$ and the x-axis from $x = 0$ to $x = \frac{\pi}{4}$.

(b) This area is rotated about the x-axis through 2π radians. Calculate the exact value of the volume formed.

2) Find the area enclosed by the two parabolas $y = 4 - x^2$ and $y = x^2 + 2x$

3) (a) Find the exact area enclosed by curve $y = e^x$, the y-axis , the x-axis and the line $x = 2$

(b) This area is rotated about the x-axis through 2π radians. Calculate the exact value of the volume formed.

4) (a) Find the exact area enclosed by curve $y = e^x$, the y-axis and the line $y = 2$.

(b) This area is rotated about the y-axis through 2π radians. Calculate the exact value of the volume formed.

5) Find the area enclosed by the two parabolas $y = x^2 - 3x - 6$ and $y = -x^2 + 5x + 18$

6) (a) Use a GDC (Graphing Display Calculator) to find the two points of intersection of the graphs of $y = \cos(0.5x)$ and $y = e^x$ in the interval $[-2\pi, 2\pi]$ correct to 4 decimal places when appropriate.

(b) Write down an expression for the area enclosed by these two graphs between those two points of intersection.

(c) Use a GDC to calculate this area correct to 3 decimal places .

7) The graph of the function $y = \ln x$ is rotated through 2π radians about the x-axis from $x = 1$ to $x = 2$.

(a) Write down an expression for the volume formed.

(b) Use a GDC to calculate this volume correct to 3 decimal places .

8) (a) The graph of $y = x^3$ is rotated through 2π radians about the x-axis from $x = 0$ to $x = 1$. Find the exact value of the volume formed.

(b) The graph of $y = x^3$ is rotated through 2π radians about the y-axis from $y = 0$ to $y = 1$. Find the exact value of the volume formed.

9) Find the exact area enclosed by $y = \sin x$ and $y = \cos 2x$ between their first two positive points of intersection.

10) (a) Sketch the graph of the function $y = 4x^4$

(b) Find the exact area enclosed by the function , the x-axis and the line $x = 2$

(c) This area is rotated through 2π radians about the x-axis. Find the exact value of the volume.

(d) This area is rotated through 2π radians about the y-axis. Find the exact value of the volume.

First Order Differential Equations

1) Variables Separable

If $\dfrac{dy}{dx} = f(x)f(y)$ then $\dfrac{1}{f(y)}\ dy = f(x)\ dx$

and $\displaystyle\int \frac{1}{f(y)}\ dy = \int f(x)\ dx$

Simple case : Solve $\dfrac{dy}{dx} = 3x^2 y$

Rearranging gives $\dfrac{1}{y}\ dy = 3x^2\ dx$

$$\int \frac{1}{y}\ dy = \int 3x^2\ dx$$

$$\ln y = x^3 + c$$

Note : you only need one constant term

Hence $y = e^{x^3 + c}$

This can be simplified to $y = A\,e^{x^3}$

Note : here $A = e^{c}$ which is just a constant

The answer $y = A\,e^{x^3}$ is called a general solution.

Example : Given the differential equation $\frac{dy}{dx} = \frac{x}{y}$

(a) Find the general solution

(b) Find the particular solution when $x = 4$ and $y = 3$

Step 1 : Separate the variables

$$y\,dy = x\,dx$$

Step 2 : Integrate both sides

$$\int y\,dy = \int x\,dx$$

$$\frac{y^2}{2} = \frac{x^2}{2} + c$$

$$y^2 = x^2 + k$$

$$y = \pm\sqrt{x^2 + k}$$

This is the general solution

Step 3 : Substitute $x = 4$ and $y = 3$ and find the value of k

$$y^2 = x^2 + k$$
$$9 = 16 + k$$
$$k = -7$$

Hence $y = \pm\sqrt{x^2 - 7}$

This is the particular solution.

Example : Given the differential equation $\tan 2x \dfrac{dy}{dx} = \sec y$

(a) Find the general solution

(b) Find the particular solution when $x = \dfrac{\pi}{4}$ and $y = \dfrac{\pi}{6}$

Step 1 : Separate the variables

$$\frac{1}{\sec y}\, dy = \frac{1}{\tan 2x}\, dx$$

$$\cos y\, dy = \cot 2x\, dx$$

Step 2 : Integrate both sides

$$\int \cos y\, dy = \int \cot 2x\, dx$$

$$\sin y = \frac{1}{2}\ln \sin 2x + c$$

$$\sin y = \ln\sqrt{\sin 2x} + c$$

$$y = \arcsin\left(\ln\sqrt{\sin 2x} + c\right)$$

This is the general solution.

Step 3 : Substitute $x = \frac{\pi}{4}$ and $y = \frac{\pi}{6}$ to find the value of c

$$\sin y = \ln\sqrt{\sin 2x} + c$$

$$\sin\frac{\pi}{6} = \ln\sqrt{\sin\frac{\pi}{2}} + c$$

$$\frac{1}{2} = \ln\sqrt{1} + c$$

$$c = \frac{1}{2}$$

Hence $y = \arcsin\left(\ln\sqrt{\sin 2x} + \frac{1}{2}\right)$

This is the particular solution

Example :

The rate of decay of a radioactive substance is proportional to the amount of the substance that has not decayed. The half-life of the radioactive substance is 10 years.

(a) Find an equation that models this decay.

(b) How long will it take for 500 grams of the substance to decay to 100 grams ?

Step 1 : Let x grams be the amount of material present that has not decayed.

Then $$\frac{dx}{dt} \propto x$$

$$\frac{dx}{dt} = k\,x$$

Step 2 : Separate the variables and integrate both sides.

$$\frac{1}{x}\,dx = k\,dt$$

$$\int \frac{1}{x}\,dx = \int k\,dt$$

$$\ln x = kt + c$$

$$x = e^{kt+c}$$

$$x = A\,e^{kt}$$

Step 3 : Find the value of A

We know the half-life is 10 years.

Hence A will become $\frac{1}{2}A$ when $t = 10$

$$\frac{1}{2}A = A\,e^{k(10)}$$

$$0.5 = e^{10k}$$

$$\ln 0.5 = \ln e^{10k}$$

$$= 10k \ln e$$

$$= 10k$$

$$\text{Hence} \quad k = \frac{\ln 0.5}{10}$$

Therefore the equation that models the decay is given by

$$x = A\,e^{\frac{\ln 0.5}{10}t}$$

Step 4 : Substitute $A = 500$ and $x = 100$

$$100 = 500\,e^{\frac{\ln 0.5}{10}t}$$

$$0.2 = e^{\frac{\ln 0.5}{10}t}$$

$$\ln 0.2 = \ln e^{\frac{\ln 0.5}{10}t}$$

$$\ln 0.2 = \frac{\ln 0.5}{10}t \ln e$$

$$\ln 0.2 = \frac{\ln 0.5}{10}t \qquad \text{since } \ln e = 1$$

$$t = \frac{10 \ln 0.2}{\ln 0.5}$$

$$t = 23.219 \text{ years}$$

2) Exact differential equations

This method uses reverse implicit differentiation to get the answer.

For example :

$x^2 \dfrac{dy}{dx} + 2xy = e^x$ can be written as $\dfrac{d}{dx}\left(x^2 y\right) = e^x$

Integrating both sides with respect to x gives

$$x^2 y = \int e^x \, dx$$

$$x^2 y = e^x + c$$

$$y = \frac{e^x + c}{x^2}$$

Example :

Given the exact differential equation $\cos x \dfrac{dy}{dx} - y \sin x = 1$

(a) Find the general solution.

(b) Find the particular solution when $x = 0$ and $y = 2$

Step 1 : Write $\cos x \dfrac{dy}{dx} - y \sin x$ as $\dfrac{d}{dx}(y \cos x)$ giving

$$\frac{d}{dx}(y \cos x) = 1$$

Step 2 : Integrate both sides with respect to x

$$y\cos x = \int 1\, dx$$

$$y\cos x = x + c$$

$$y = \frac{x+c}{\cos x}$$

This is the general solution.

Step 3 : Substitute $x = 0$ and $y = 2$ to find the value of c

$$2\cos 0 = 0 + c$$

$$c = 2$$

Hence $\quad y\cos x = x + 2$

$$y = \frac{x+2}{\cos x}$$

This is the particular solution

Example : Given the exact differential equation $\ln x\frac{dy}{dx} + \frac{y}{x} = \cos x$

(a) Find the general solution.

(b) Find the particular solution when $x = \frac{\pi}{6}$ and $y = 0$

Step 1 : Write $\ln x\frac{dy}{dx} + \frac{y}{x}$ as $\frac{d}{dx}(y\ln x)$ giving

$$\frac{d}{dx}(y\ln x) = \cos x$$

Step 2 : Integrate both sides with respect to x

$$y\ln x = \int \cos x \, dx$$

$$= \sin x + c$$

$$y = \frac{\sin x + c}{\ln x}$$

This is the general solution.

Step 3 : Substitute $x = \frac{\pi}{6}$ and $y = 0$ to find the value of c

$$0 = \frac{\sin\frac{\pi}{6} + c}{\ln\frac{\pi}{6}}$$

$$0 = \sin\frac{\pi}{6} + c$$

$$0 = \frac{1}{2} + c$$

$$c = -\frac{1}{2}$$

Hence $$y = \frac{\sin x - \frac{1}{2}}{\ln x}$$

This is the particular solution

3) Integrating factor

This method is used when you have differential equations of the type

$$\frac{dy}{dx} + P(x)y = Q(x)$$

The integrating factor will be $e^{\int P(x)\,dx}$

For example to solve $\dfrac{dy}{dx} + 2y = x$

the integrating factor will be $e^{\int 2\,dx}$ or e^{2x}

Then $\dfrac{dy}{dx} + 2y = x$ becomes $e^{2x}\dfrac{dy}{dx} + 2ye^{2x} = xe^{2x}$

(just multiply both sides by the integrating factor).

This always results in the left hand side being an exact integral.

$$\frac{d}{dx}\left(ye^{2x}\right) = xe^{2x}$$

Integrating both sides gives

$$ye^{2x} = \int xe^{2x}dx$$

$$ye^{2x} = x.\tfrac{1}{2}e^{2x} - \int \tfrac{1}{2}e^{2x}.\,1\,dx \quad \text{(using integration by parts)}$$

$$ye^{2x} = \tfrac{1}{2}xe^{2x} - \tfrac{1}{2}\int e^{2x}\,dx$$

$$ye^{2x} = \tfrac{1}{2}xe^{2x} - \tfrac{1}{4}e^{2x} + c$$

$$y = \tfrac{1}{2}x - \tfrac{1}{4} + \frac{c}{e^{2x}}$$

Example :

Given the differential equation $\frac{dy}{dx} + y\cot x = 4\cos x$

(a) Find the general solution

(b) Find the particular solution when $x = \frac{\pi}{6}$ and $y = 1$

Step 1 : Find the integrating factor

$$e^{\int \cot x \, dx} = e^{\ln \sin x}$$

$$= \sin x$$

Step 2 : Multiply both sides by the integrating factor

$$\sin x\left(\frac{dy}{dx} + y\cot x\right) = (4\cos x)\sin x$$

$$\sin x\frac{dy}{dx} + y\cos x = 2(2\sin x\cos x)$$

$$\sin x\frac{dy}{dx} + y\cos x = 2\sin 2x$$

Step 3 : Write $\sin x\frac{dy}{dx} + y\cos x$ as $\frac{d}{dx}(y\sin x)$

and then integrate both sides with respect to x

$$\frac{d}{dx}(y\sin x) = 2\sin 2x$$

$$y\sin x = \int 2\sin 2x \, dx$$

$$y \sin x = -\cos 2x + c$$

$$y = -\frac{\cos 2x}{\sin x} + \frac{c}{\sin x}$$

$$y = c \csc x - \cos 2x \csc x$$

This is the general solution

Step 4 : Substitute $x = \frac{\pi}{6}$ and $y = 1$ to find the value of c

$$1 = c \csc\frac{\pi}{6} - \cos\frac{\pi}{3}\csc\frac{\pi}{6}$$

$$1 = c\,(2) - \frac{1}{2}(2)$$

$$1 = 2c - 1$$

$$c = 1$$

Hence $y = \csc x - \cos 2x \csc x$ is the particular solution.

4) Homogeneous differential equations

This method is used when you have differential equations of the type

$$\frac{dy}{dx} = f\left(\frac{y}{x}\right)$$

For example $\quad x^2\frac{dy}{dx} = y^2 + xy + 4x^2$

This can be written as $\quad \dfrac{dy}{dx} = \dfrac{y^2 + xy + 4x^2}{x^2}$

$$= \frac{y^2}{x^2} + \frac{xy}{x^2} + \frac{4x^2}{x^2}$$

$$= \left(\frac{y}{x}\right)^2 + \frac{y}{x} + 4$$

This is now in the form $\quad \dfrac{dy}{dx} = f\left(\dfrac{y}{x}\right)$

This type is solved by letting $\; y = ux \;$ where u is a function of x

When $\; y = ux \;$ we have $\; \dfrac{dy}{dx} = u + x\dfrac{du}{dx}$

Hence $\; \dfrac{dy}{dx} = \left(\dfrac{y}{x}\right)^2 + \dfrac{y}{x} + 4 \;$ becomes

$$u + x\frac{du}{dx} = u^2 + u + 4$$

$$x\frac{du}{dx} = u^2 + 4$$

This is now a differential equation of the type "variables separable"

$$x\frac{du}{dx} = u^2+4 \quad \Rightarrow \quad \frac{1}{u^2+4}\,du = \frac{1}{x}\,dx$$

$$\int\frac{1}{u^2+4}\,du = \int\frac{1}{x}\,dx$$

$$\int\frac{1}{2^2+u^2}\,du = \int\frac{1}{x}\,dx$$

$$\frac{1}{2}\arctan\frac{u}{2} = \ln x + c$$

$$\frac{1}{2}\arctan\frac{u}{2} = \ln Ax$$

$$\arctan\frac{u}{2} = 2\ln Ax$$

$$\frac{u}{2} = \tan\left(2\ln Ax\right)$$

$$u = 2\tan\left(2\ln Ax\right)$$

$$\frac{y}{x} = 2\tan\left(2\ln Ax\right)$$

$$y = 2x\tan\left(2\ln Ax\right)$$

This is the general solution.

Example : (a) Show that the differential equation $x\frac{dy}{dx} - y = 2x$ is homogeneous .

(b) Find the general solution.

(c) Find the particular solution when $x = 1$ and $y = 0$

Step 1 : Re-arrange the differential equation

$$x\frac{dy}{dx} = 2x + y$$

$$\frac{dy}{dx} = \frac{2x + y}{x}$$

$$= 2 + \frac{y}{x}$$

$$= f\left(\frac{y}{x}\right) \quad \text{hence homogeneous}$$

Step 2 : Let $y = ux$ and $\frac{dy}{dx} = u + x\frac{du}{dx}$

$$\frac{dy}{dx} = 2 + \frac{y}{x} \quad \text{gives} \quad u + x\frac{du}{dx} = 2 + u$$

$$x\frac{du}{dx} = 2$$

$$1\ du = \frac{2}{x}\ dx$$

$$\int 1\ du = \int \frac{2}{x}\ dx$$

$$u = 2\ln x + c$$

$$u = \ln Ax^2$$

$$\frac{y}{x} = \ln Ax^2$$

$$y = x\ln Ax^2$$

This is the general solution.

Step 3 : Substitute $x = 1$ and $y = 0$ to find the value of A

$$y = x\ln Ax^2$$

$$0 = \ln A$$

$$A = 1$$

Hence $y = x\ln x^2$ is the particular solution.

Euler's Method

There are two ways to obtain particular solutions to differential equations :

1) Accurately : using integration
2) Not-so-accurately : using iteration (Euler's Method)

Consider the following graph :

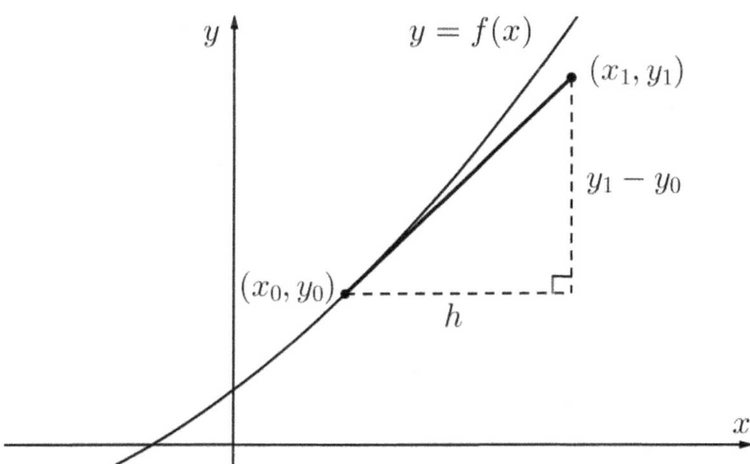

The graph of $y = f(x)$ represents the solution to a differential equation.

The coordinate (x_0, y_0) is a given point on the solution curve.

The coordinate (x_1, y_1) is a point that is almost on the solution curve.

By making h smaller and smaller we can make (x_1, y_1) closer and closer to the curve.

The line joining (x_0, y_0) and (x_1, y_1) is a tangent to the curve at the point (x_0, y_0). Hence its gradient is given by $f'(x_0)$

This means that $f'(x_0) = \dfrac{y_1 - y_0}{h}$

Rearranging gives $y_1 = y_0 + hf'(x_0)$

This is the basis of Euler's iteration formula.

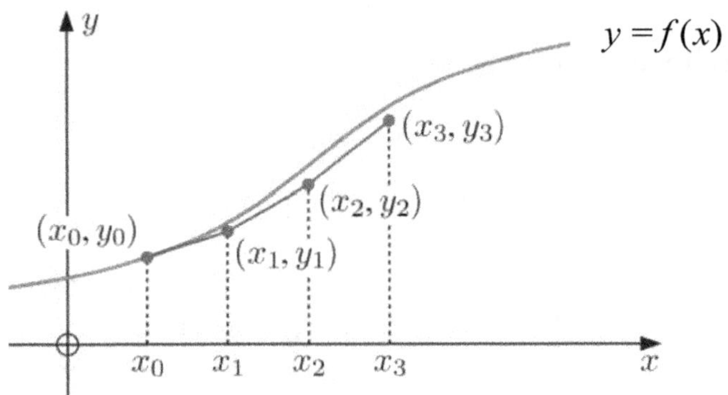

$$y_1 = y_0 + hf'(x_0)$$

$$y_2 = y_1 + hf'(x_1)$$

$$y_3 = y_2 + hf'(x_2)$$

$$y_4 = y_3 + hf'(x_3)$$

etc.

In general

$y_{n+1} = y_n + hf'(x_n)$ where h = step length

Since we are dealing with a differential equation that already contains $\frac{dy}{dx}$ we do not actually need to perform any differentiation to use Euler's formula.

The differential equation that we are trying to solve will have the form $\frac{dy}{dx} = f(x, y)$ where $f(x, y)$ represents a function containing terms in x and y.

For example $\frac{dy}{dx} = xy - 1$

Since $f(x, y) = \frac{dy}{dx}$ we use the notation $f(x_n, y_n)$ to indicate $\frac{dy}{dx}$ at the point (x_n, y_n)

In other words $f(x_n, y_n)$ = gradient of the curve at the point (x_n, y_n)

This gives the Euler's iteration formula that you will find on the formula sheet

$$y_{n+1} = y_n + hf(x_n, y_n)$$

$$x_{n+1} = x_n + h$$

For example , consider the differential equation $\frac{dy}{dx} = xy - 1$ with initial point (0 , 3).

This means that the point (0 , 3) lies on the solution curve $y = f(x)$.

We can use Euler's Method to estimate the value of y when $x = 0.7$

Here $(x_0, y_0) = (0 , 3)$ and $f(x, y) = xy - 1$

We can use step size $h = 0.1$ to get a reasonable answer
(smaller values for h will , of course, give a more accurate result)

Since $x_{n+1} = x_n + h$ we know $x_1 = x_0 + 0.1$

$$= 0.1$$

Since $y_1 = y_0 + hf(x_0, y_0)$ we know $y_1 = 3 + 0.1f(x_0, y_0)$

$$= 3 + 0.1\left[(x_0)(y_0) - 1\right]$$
$$= 3 + 0.1\left[(0)(3) - 1\right]$$
$$= 2.9$$

This process is repeated until we reach $x = 0.7$

The following table shows the required calculations :

n	x_n	y_n
0	0	3
1	0.1	$y_1 = y_0 + h[f(x_0, y_0)]$ $= 3 + 0.1[(0)(3) - 1]$ $= 2.9$
2	0.2	$y_2 = y_1 + h[f(x_1, y_1)]$ $= 2.9 + 0.1[(0.1)(2.9) - 1]$ $= 2.829$
3	0.3	$y_3 = y_2 + h[f(x_2, y_2)]$ $= 2.829 + 0.1[(0.2)(2.829) - 1]$ $= 2.78558$
4	0.4	$y_4 = y_3 + h[f(x_3, y_3)]$ $= 2.78558 + 0.1[(0.3)(2.78558) - 1]$ $= 2.7691474$

5	0.5	$y_5 = y_4 + h\left[f(x_4, y_4)\right]$ $= 2.7691474 + 0.1\left[(0.4)(2.7691474) - 1\right]$ $= 2.7799133$
6	0.6	$y_6 = y_5 + h\left[f(x_5, y_5)\right]$ $= 2.7799133 + 0.1\left[(0.5)(2.7799133) - 1\right]$ $= 2.8189090$
7	0.7	$y_7 = y_6 + h\left[f(x_6, y_6)\right]$ $= 2.8189090 + 0.1\left[(0.6)(2.8189090) - 1\right]$ $= 2.888043$

Hence when $x = 0.7$ the value of y is 2.888 (correct to 3 decimal places)

An Excel spreadsheet can be used to check your work :

C3 f_x =C2+0.1*(B2*C2-1)

	A	B	C	D
1	n	Xn	Yn	
2	0	0	3	
3	1	0.1	2.9	
4	2	0.2	2.829	
5	3	0.3	2.78558	
6	4	0.4	2.7691474	
7	5	0.5	2.779913296	
8	6	0.6	2.818908961	
9	7	0.7	2.888043498	
10				

The formula required to calculate y_n is shown above column C.

For x_n use the Excel formula shown below :

B3		fx =B2+0.1

	A	B	C
1	n	Xn	Yn
2	0	0	3
3	1	0.1	2.9
4	2	0.2	2.829
5	3	0.3	2.78558
6	4	0.4	2.7691474

Example :

The differential equation $\frac{dy}{dx} = 2xy - y$ has a solution given by $y = f(x)$.

The point (2 ,3) lies on the solution curve.

(a) Use Euler's method with step length 0.01 to find an approximate value of y when $x = 2.05$
Give your answer correct to 4 decimal places.

(b) Solve the differential equation $\frac{dy}{dx} = 2xy - y$ **by using integration** and then find the particular solution that contains the point (2 ,3).

(c) Hence find the value of y (correct to 4 decimal places) when $x = 2.05$

(d) Calculate the percentage error in Eulers method for this particular problem (correct to 3 decimal places)

Step 1 : Let $(x_0, y_0) = (2, 3)$

$$\begin{aligned} \text{Then } x_1 &= x_0 + 0.01 \\ &= 2.01 \end{aligned}$$

$$\begin{aligned} \text{and } y_1 &= y_0 + 0.01\left[f(x_0, y_0)\right] \\ &= y_0 + 0.01\left[2(x_0)(y_0) - y_0\right] \\ &= 3 + 0.01\left[2(2)(3) - 3\right] \\ &= 3.09 \end{aligned}$$

Step 2 : Continue the iteration

$$\begin{aligned} x_2 &= x_1 + 0.01 \\ &= 2.02 \end{aligned}$$

$$\begin{aligned} y_2 &= y_1 + 0.01\left[f(x_1, y_1)\right] \\ &= y_1 + 0.01\left[2(x_1)(y_1) - y_1\right] \\ &= 3.09 + 0.01\left[2(2.01)(3.09) - 3.09\right] \\ &= 3.183318 \end{aligned}$$

$$\begin{aligned} x_3 &= x_2 + 0.01 \\ &= 2.03 \end{aligned}$$

$$\begin{aligned} y_3 &= y_2 + 0.01\left[f(x_2, y_2)\right] \\ &= y_2 + 0.01\left[2(x_2)(y_2) - y_2\right] \\ &= 3.183318 + 0.01\left[2(2.02)(3.183318) - 3.183318\right] \\ &= 3.280090867 \\ &= 3.280091 \text{ (correct to 6 decimal places)} \end{aligned}$$

$x_4 = 2.04$

$y_4 = 3.280091 + 0.01\left[2(2.03)(3.280091) - 3.280091\right]$

$= 3.380462$

$x_5 = 2.05$

$y_5 = 3.380462 + 0.01\left[2(2.04)(3.380462) - 3.380462\right]$

$= 3.484580$

Hence when $x_5 = 2.05$ $y_5 = 3.4846$ (correct to 4 decimal places)

Step 3 : Solve the differential equation by separating the variables

$$\frac{dy}{dx} = 2xy - y$$

$$\frac{dy}{dx} = y(2x-1)$$

$$\frac{1}{y}\,dy = (2x-1)\,dx$$

$$\int \frac{1}{y}\,dy = \int (2x-1)\,dx$$

$$\ln y = x^2 - x + c$$

Substitute (2,3) : $\ln 3 = 2^2 - 2 + c$

$$c = \ln 3 - 2$$

$$\ln y = x^2 - x + \ln 3 - 2$$

$$y = e^{x^2 - x + \ln 3 - 2}$$

Step 4 : Let $x = 2.05$

Then $y = e^{2.05^2 - 2.05 + \ln 3 - 2}$

$= 3.4942$(correct to 4 decimal places)

Step 5 : Calculate percentage error

$$\text{Actual error} = 3.4942 - 3.4846 = 0.0096$$

$$\%\ \text{error} = \frac{0.0096}{3.4942} \times 100\ \%$$

$= 0.2136\ \%$

$= 0.214\ \%$ (correct to 3 decimal places)

Exercise 8

1) Given the differential equation $y\frac{dy}{dx} = 4\cos 2x$

   (a) Find the general solution.

   (b) Find the particular solution if $y = 1$ when $x = \frac{\pi}{12}$

2) Solve $\left(x^2+1\right)\frac{dy}{dx} = xy$

3) Given the exact differential equation $x\cos y\frac{dy}{dx} + \sin y = 1$

   (a) Find the general solution.

   (b) Find the particular solution when $x = 2$ and $y = \frac{\pi}{6}$

4) Find the general solution to $x^2e^y\frac{dy}{dx} + 2xe^y = 3$

5) Given the differential equation $\frac{dy}{dx} + y\cot x = \cos x$

   (a) Find the integrating factor

   (b) Hence find the general solution

   (c) Find the particular solution when $y = 1$ and $x = \frac{\pi}{6}$

6) Solve $\frac{dy}{dx} + \frac{y}{x} = 3x$

7) Given $\frac{dy}{dx}+3x^2y=6x^2$ find the general solution by

(a) Separating the variables

(b) Using an integrating factor

(c) Show that your two answers are identical

8) The rate of growth of a colony of bacteria is directly proportional to the number of bacteria present.

(a) If the initial number of bacteria present is 50 then find an equation that models this growth.

(b) After 10 hours number of bacteria present is 500. Find the number of bacteria present after 12 hours.

(c) How long will it take for the bacteria to number one million ?

9) Given $\frac{dy}{dx}+y=e^x$ has a solution $y=f(x)$ such that $f(3)=2$.

(a) Use Euler's method with a step length of 0.1 to approximate the value of y when $x=3.3$ Show all working and give your final answer correct to 3 decimal places.

(b) Use an integrating factor to solve the differential equation.

(c) Hence find the actual value of $f(3.3)$ correct to 3 decimal places.

10) The differential equation $\frac{dy}{dx}=x^2+y^2$ has the solution $y=f(x)$ where $f(0)=1$. Use Euler's method with a step length of 0.1 to approximate the value of y when $x=0.4$ Show all working and give your final answer correct to 3 decimal places.

Maclaurin Series

$$f(x) = f(0) + xf'(0) + \frac{x^2}{2!}f''(0) + \frac{x^3}{3!}f'''(0) + ...$$

This is a power series that provides a way to expand out a variety of functions and then calculate an approximate value of the function (usually values close to zero)

For example : $f(x) = e^x$

$$f(x) = e^x \quad \Rightarrow \quad f(0) = 1$$

$$f'(x) = e^x \quad \Rightarrow \quad f'(0) = 1$$

$$f''(x) = e^x \quad \Rightarrow \quad f''(0) = 1$$

$$f'''(x) = e^x \quad \Rightarrow \quad f'''(0) = 1$$

$$f(x) = f(0) + xf'(0) + \frac{x^2}{2!}f''(0) + \frac{x^3}{3!}f'''(0) + ...$$

$$e^x = 1 + x\,(1) + \frac{x^2}{2!}(1) + \frac{x^3}{3!}(1) + ...$$

$$e^x = 1 + x + \frac{x^2}{2!} + \frac{x^3}{3!} + ...$$

This expansion is valid for all x.
However it will converge faster if x is small.

For example

$$e^{0.2} = 1 + 0.2 + \frac{(0.2)^2}{2!} + \frac{(0.2)^3}{3!} + ...$$

$$= 1 + 0.2 + 0.02 + 0.00133 + ...$$

$$= 1.\,22133 \ldots\ldots$$

This is already very close to the actual value of 1.221402 ....

Whereas $$e^2 = 1 + 2 + \frac{(2)^2}{2!} + \frac{(2)^3}{3!} + ...$$

$$= 1 + 2 + 2 + 1.33333 + ...$$

$$= 6.333333$$

This needs more terms in order to approach the actual value of 7.389056 .....

The Maclaurin series for e^x can also be used to find expansions for functions such as e^{x^2}

$$e^x = 1 + x + \frac{x^2}{2!} + \frac{x^3}{3!} + ...$$

$$e^{x^2} = 1 + x^2 + \frac{\left(x^2\right)^2}{2!} + \frac{\left(x^2\right)^3}{3!} + ...$$

$$e^{x^2} = 1 + x^2 + \frac{x^4}{2!} + \frac{x^6}{3!} + ...$$

Example :

(a) Find the first four non-zero terms of the Maclaurin series for $\sin x$. Hence find an approximation for $\sin 0.8$

(b) State the number of decimal places to which the approximation is accurate.

Step 1 : Let $f(x) = \sin x$ and find $f(0)$

$$f(x) = \sin x \quad \Rightarrow \quad f(0) = 0$$

Step 2 : Differentiate

$$f'(x) = \cos x \quad \Rightarrow \quad f'(0) = 1$$

$$f''(x) = -\sin x \quad \Rightarrow \quad f''(0) = 0$$

$$f'''(x) = -\cos x \quad \Rightarrow \quad f'''(0) = -1$$

$$f^{iv}(x) = \sin x \quad \Rightarrow \quad f^{iv}(0) = 0$$

$$f^{v}(x) = \cos x \quad \Rightarrow \quad f^{v}(0) = 1$$

$$f^{vi}(x) = -\sin x \quad \Rightarrow \quad f^{vi}(0) = 0$$

$$f^{vii}(x) = -\cos x \Rightarrow f^{vii}(0) = -1$$

Step 3 : Substitute into the Maclaurin series

$$f(x) = f(0) + xf'(0) + \frac{x^2}{2!}f''(0) + \frac{x^3}{3!}f'''(0) + \ldots\ldots.$$

$$\sin x = 0 + x\,(1) + \frac{x^2}{2!}(0) + \frac{x^3}{3!}(-1) + \frac{x^4}{4!}(0) + \frac{x^5}{5!}(1)$$

$$+ \frac{x^6}{6!}(0) + \frac{x^7}{7!}(-1) + \ldots\ldots$$

$$\sin x = x - \frac{x^3}{3!} + \frac{x^5}{5!} - \frac{x^7}{7!} + \ldots\ldots.$$

Step 4 : Substitute $x = 0.8$

$$\sin 0.8 = 0.8 - \frac{(0.8)^3}{3!} + \frac{(0.8)^5}{5!} - \frac{(0.8)^7}{7!} + \ldots\ldots.$$

$$= 0.7173557232$$

Step 5 : Compare your result to the actual value of sin 0.8

$\sin 0.8 = 0.7173560909$

The two results are identical up to 6 decimal places.

The Maclaurin series can be used to obtain the binomial expansion of $(1+x)^n$

If $f(x)=(1+x)^n$ then $f(0)=1$

Differentiating $f(x)=(1+x)^n$ gives

$$f'(x)=n(1+x)^{n-1} \quad \Rightarrow \quad f'(0)=n$$

$$f''(x)=n(n-1)(1+x)^{n-2} \quad \Rightarrow \quad f''(0)=n(n-1)$$

$$f'''(x)=n(n-1)(n-2)(1+x)^{n-3} \quad \Rightarrow \quad f'''(0)=n(n-1)(n-2)$$

Substituting into

$$f(x) = f(0) + x f'(0) + \frac{x^2}{2!} f''(0) + \frac{x^3}{3!} f'''(0) + \ldots\ldots. \quad \text{gives}$$

$$(1+x)^n = 1+x\,(n)+\frac{x^2}{2!}(n)(n-1)+\frac{x^3}{3!}(n)(n-1)(n-2)+\ldots..$$

$$(1+x)^n = 1+nx+(n)(n-1)\frac{x^2}{2!}+(n)(n-1)(n-2)\frac{x^3}{3!}+\ldots..$$

$$(1+x)^n = 1+nx+\frac{(n)(n-1)}{2!}x^2+\frac{(n)(n-1)(n-2)}{3!}x^3+\ldots..$$

Here if n is a positive integer then the series is finite and valid for all x.

However , if n is negative or fractional then the series is infinite and only valid for $-1<x<1$

Common Maclaurin Series

$$e^x = 1 + x + \frac{x^2}{2!} + \frac{x^3}{3!} + \ldots\ldots. \qquad x \in \mathbb{R}$$

$$\sin x = x - \frac{x^3}{3!} + \frac{x^5}{5!} - \frac{x^7}{7!} + \ldots\ldots. \qquad x \in \mathbb{R}$$

$$\cos x = 1 - \frac{x^2}{2!} + \frac{x^4}{4!} - \frac{x^6}{6!} + \ldots\ldots. \qquad x \in \mathbb{R}$$

$$\ln(1+x) = x - \frac{x^2}{2} + \frac{x^3}{3} - \frac{x^4}{4} + \ldots\ldots \qquad -1 < x \le 1$$

$$\arctan x = x - \frac{x^3}{3} + \frac{x^5}{5} - \frac{x^7}{7} + \ldots\ldots.. \qquad -1 \le x \le 1$$

Example : Use $\ln(1+x) = x - \frac{x^2}{2} + \frac{x^3}{3} - \frac{x^4}{4}$

to find the Maclaurin series for $\frac{1}{1+x}$

Step 1 : Notice that $\frac{d}{dx}\left[\ln(1+x)\right] = \frac{1}{1+x}$

Step 2 : Differentiate both sides of the series

$$\frac{d}{dx}\left[\ln(1+x)\right] = \frac{d}{dx}\left[x - \frac{x^2}{2} + \frac{x^3}{3} - \frac{x^4}{4} + \ldots\ldots\right]$$

$$\frac{1}{1+x} = 1 - \frac{2x}{2} + \frac{3x^2}{3} - \frac{4x^3}{4} + \ldots\ldots$$

$$\frac{1}{1+x} = 1 - x + x^2 - x^3 + \ldots\ldots$$

Example : Use the Maclaurin series for e^x and $\sin x$ to find a power series for $e^x \sin x$ up to the term containing x^5.

$$e^x \sin x = \left(1 + x + \frac{x^2}{2!} + \frac{x^3}{3!} + \ldots\right)\left(x - \frac{x^3}{3!} + \frac{x^5}{5!} - \frac{x^7}{7!} + \ldots\right)$$

$$= x - \frac{x^3}{3!} + \frac{x^5}{5!} + x^2 - \frac{x^4}{3!} + \frac{x^3}{2!} - \frac{x^5}{2!3!} + \frac{x^4}{3!}$$

ignoring powers of x greater than 5

$$= x + x^2 - \frac{x^3}{6} + \frac{x^3}{2} + \frac{x^5}{120} - \frac{x^5}{12}$$

$$= x + x^2 + \frac{x^3}{3} - \frac{3x^5}{40}$$

Maclaurin Series and Differential Equations

A Maclaurin series can also be developed from a differential equation.

For example , consider the differential equation $\dfrac{dy}{dx} = ye^x$ which has solution $y = f(x)$ and the point $(0,1)$ on the solution curve.

We can use this information to find the Maclaurin expansion of y.

We know $f(0) = 1$ from the given point on the solution curve.

Since $\dfrac{dy}{dx} = ye^x$ we have $f'(x) = ye^x$

$$f'(0) = (1)e^0$$

$$f'(0) = 1$$

To find $\dfrac{d^2y}{dx^2}$ we differentiate both sides of $\dfrac{dy}{dx} = ye^x$

$$\frac{d}{dx}\left[\frac{dy}{dx}\right] = \frac{d}{dx}\left[ye^x\right]$$

$$\frac{d^2y}{dx^2} = ye^x + e^x\frac{dy}{dx}$$

$$f''(x) = ye^x + e^x\frac{dy}{dx}$$

$$f''(0) = (1)e^0 + e^0(1)$$

$$f''(0) = 2$$

To find $\dfrac{d^3y}{dx^3}$ we differentiate both sides of $\dfrac{d^2y}{dx^2} = ye^x + e^x\dfrac{dy}{dx}$

$$\frac{d}{dx}\left[\frac{d^2y}{dx^2}\right] = \frac{d}{dx}\left[ye^x + e^x\frac{dy}{dx}\right]$$

$$\frac{d^3y}{dx^3} = ye^x + e^x\frac{dy}{dx} + e^x\frac{d^2y}{dx^2} + e^x\frac{dy}{dx}$$

$$f'''(x) = ye^x + e^x\frac{dy}{dx} + e^x\frac{d^2y}{dx^2} + e^x\frac{dy}{dx}$$

$$f'''(0) = (1)e^0 + e^0(1) + e^0(2) + e^0(1)$$

$$f'''(0) = 5$$

These results can now be substituted into the Maclaurin series

$$f(x) = f(0) + xf'(0) + \frac{x^2}{2!}f''(0) + \frac{x^3}{3!}f'''(0) + \ldots\ldots.$$

$$f(x) = 1 + x\ (1) + \frac{x^2}{2!}(2) + \frac{x^3}{3!}(5) + \ldots\ldots.$$

$$f(x) = 1 + x + x^2 + \frac{5x^3}{6} + \ldots\ldots.$$

Example :

Given the differential equation $\cos x \frac{dy}{dx} + y \sin x = 1$

has the solution $y = f(x)$ and the point $(0,2)$ on its solution curve.

(a) Find the first three terms in the Maclaurin expansion of y

(b) Solve the differential equation analytically

(c) Show that your two answers are consistent

Step 1 : $f(0) = 2$

Step 2 : Substitute $(0,2)$ into $\cos x \frac{dy}{dx} + y \sin x = 1$ to find $f'(0)$

$$\cos 0 \frac{dy}{dx} + 2 \sin 0 = 1$$

$$\frac{dy}{dx} = 1$$

Hence $f'(0) = 1$

Step 3 : Use implicit differentiation to find an expression involving $\frac{d^2 y}{dx^2}$

$$\frac{d}{dx}\left(\cos x \frac{dy}{dx} + y \sin x \right) = \frac{d}{dx}(1)$$

$$\cos x \frac{d^2 y}{dx^2} + \frac{dy}{dx}(-\sin x) + y \cos x + \frac{dy}{dx} \sin x = 0$$

Step 4 : Substitute (0,2) and $\dfrac{dy}{dx}=1$ into

$$\cos x\frac{d^2y}{dx^2}+\frac{dy}{dx}(-\sin x)+y\cos x+\frac{dy}{dx}\sin x=0 \text{ to find } f''(0)$$

$$\cos 0\frac{d^2y}{dx^2}+(1)(-\sin 0)+2\cos 0+(1)\sin 0=0$$

$$\frac{d^2y}{dx^2}+2 = 0$$

$$\frac{d^2y}{dx^2} = -2$$

Hence $f''(0) = -2$

Step 5 : Substitute values into

$$f(x) = f(0) + xf'(0) + \frac{x^2}{2!}f''(0) + \frac{x^3}{3!}f'''(0) + \ldots\ldots.$$

$$y = 2 + x(1) + \frac{x^2}{2!}(-2) + \ldots\ldots.$$

$$y = 2 + x - x^2$$

Step 6 : Solve $\cos x\dfrac{dy}{dx} + y\sin x=1$

$$\frac{dy}{dx} + y\frac{\sin x}{\cos x}=\frac{1}{\cos x}$$

$$\frac{dy}{dx} + y\tan x=\sec x$$

$$\text{Integrating factor} = e^{\int \tan x\,dx}$$

$$= e^{\ln\sec x}$$

$$= \sec x$$

Hence $\dfrac{dy}{dx} + y\tan x = \sec x \quad \rightarrow \quad \sec x\dfrac{dy}{dx} + y\sec x\tan x = \sec^2 x$

$$\frac{d}{dx}(y\sec x) = \sec^2 x$$

$$y\sec x = \int \sec^2 x\,dx$$

$$y\sec x = \tan x + c$$

Substitute (0,2) to find c

$$2\sec 0 = \tan 0 + c$$

$$c = 2$$

Hence $\quad y\sec x = \tan x + 2$

$$y = \frac{\tan x + 2}{\sec x}$$

$$y = \sin x + 2\cos x$$

Step 7 : Use Maclaurin to expand out the result $\quad y = \sin x + 2\cos x$

$$y = \left(x - \frac{x^3}{3!} + \frac{x^5}{5!} + ...\right) + 2\left(1 - \frac{x^2}{2!} + \frac{x^4}{4!} - ....\right)$$

$$y = x - \frac{x^3}{3!} + 2\left(1 - \frac{x^2}{2!}\right)$$

$$y = x - \frac{x^3}{3!} + 2 - x^2$$

$y = 2 + x - x^2 \quad$ ignoring powers higher than 2

This is the same as the result in part (a)

Exercise 9

1) (a) By differentiating $\cos x$ a suitable number of times , obtain the first four non-zero terms of the Maclaurin series for $\cos x$.

   (b) Use your answer to part (a) to find an approximation for $\cos 0.9$

   (c) State the number of decimal places to which the approximation is accurate.

2) By differentiating $\arctan x$ a suitable number of times , obtain the first three non-zero terms of the Maclaurin series for $\arctan x$.

3) Explain why it is not possible to use Maclaurin series to expand $\ln x$

4) Use $\arctan x = x - \frac{x^3}{3} + \frac{x^5}{5} - \frac{x^7}{7} + \ldots\ldots$

   to find the Maclaurin series for $\frac{1}{1+x^2}$

5) Use a known Maclaurin series to find the first four terms in the expansion of $\sin x^2$

6) Use the Maclaurin series for e^x and $\cos x$ to find a power series for $e^x \cos x$ up to the term containing x^4

7) Given $f(x) = \ln(1 - \sin x)$

(a) Show that $f''(x) = \dfrac{-1}{1 - \sin x}$

(b) Determine the Maclaurin series for $f(x)$ as far as the term in x^4

8) The differential equation $\dfrac{dy}{dx} = 2xy$ has solution $y = f(x)$.
The point $(0,1)$ lies on the solution curve.

(a) Find the first three non-zero terms in the Maclaurin expansion of y.

(b) Solve the differential equation giving the particular solution at $(0,1)$ in the form $y = f(x)$

(c) Show that your two answers are consistent.

9) Given the differential equation $\dfrac{dy}{dx} = y\ln(x+1)$ and $y(0) = 1$

Find the first three non-zero terms in the Maclaurin expansion of y

10) Given the differential equation $\dfrac{dy}{dx} = 2x - y$ and $y(0) = 1$

(a) Find the first four terms in the Maclaurin expansion of y

(b) Solve the differential equation analytically.

(c) Show that your two answers are consistent.

Answers :

Exercise 1

1) (a) $45x^4$ (b) $3x^2+2x+1$ (c) $\frac{4}{\sqrt{x}}$

(d) $\frac{-6}{x^4}$ (e) $\frac{-3}{\sqrt{x^3}}$ (f) $8x+4$

2) (a) $\frac{-1}{x^2}$ (b) $\frac{10}{\sqrt[3]{x}}$ (c) π (d) $3+\frac{2}{x^2}$

3) (a) 12 (b) $y=12x-7$

4) (a) $-\frac{1}{4}$ (b) $y=4x-\frac{31}{4}$

5) (a) $v=12t^2-6t$ (b) 36 m/s (c) $0\,,\ \frac{1}{2}$ sec

(d) $24t-6$, acceleration

6) −24

7) 200π cm^3/sec

11) $\frac{2}{25}$ cm/sec

12) $\frac{1}{2\pi}$ cm/sec

13) $\frac{2}{25\pi}$ cm/sec

Exercise 2

1) (a) $72(3x+5)^5$ (b) $45x^2(3x^3+4)^4$ (c) $\frac{12}{\sqrt{3x-5}}$

(d) $\frac{16}{(1-4x)^3}$ (e) $\frac{-5}{(5x+2)^2}$ (f) $\frac{2}{\sqrt[3]{(6x+1)^4}}$

2) (a) $2x(6x-1)^4(21x-1)$ (b) $\frac{3x+2}{\sqrt{x+1}}$

(c) $\frac{-6}{(2x-1)^2}$ (d) $\frac{x-1}{\sqrt{(2x-1)^3}}$

3) (a) -2 (b) $y=-2x+1$

4) (a) $\frac{2}{3}$ (b) $y=-\frac{3}{2}x+\frac{15}{2}$

5) (a) $(x+1)^2(x+5)^3(7x+19)$ (b) $x=-1,-5,-\frac{19}{7}$

6) $60\sqrt{4x+1}$

7) (a) $6x^2+30x-36$, $x=-6$, 1 (b) -6 max , 1 min

(c) $12x+30$ (e) $x=-\frac{5}{2}$ (f) $x<-6$, $x>1$

(g) $-6<x<1$ (h) $x>-\frac{5}{2}$ (i) $x<-\frac{5}{2}$

8)

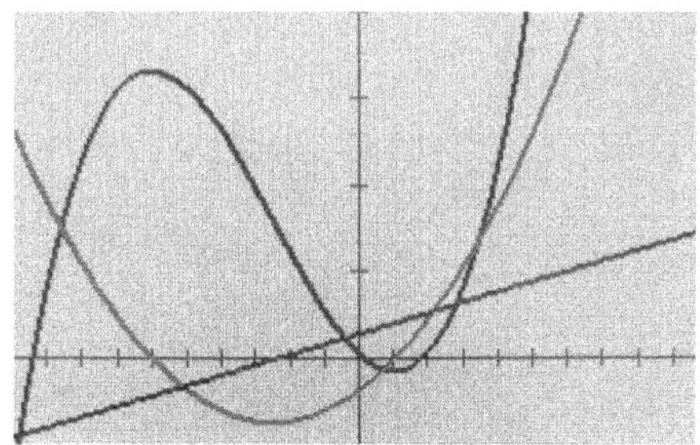

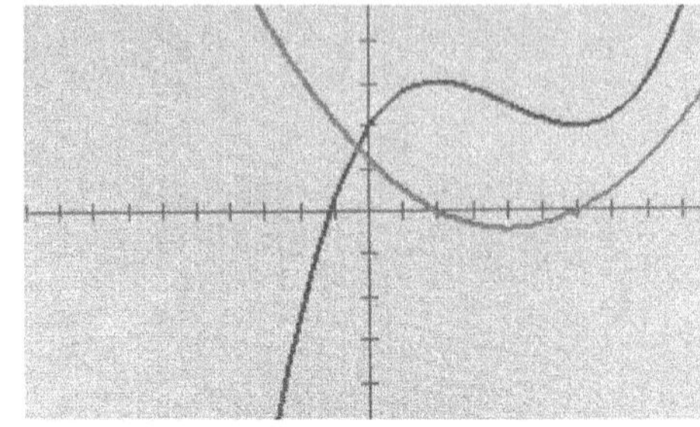

10)

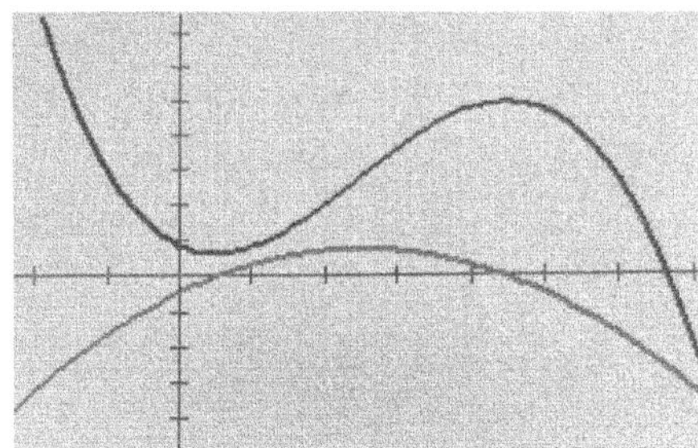

11) 648 cm^2

12) (a) $4x^3 - 44x^2 + 120x$ (b) $12x^2 - 88x + 120$ (c) 1.81 cm

13) (a) $V = \frac{1}{3}\pi r\sqrt{100 - r^4}$ (b) $\sqrt[4]{\frac{100}{3}}$

14) (a) 27

15) (b) $\sqrt{b^2 - b + 9}$ (c) 2.958

Exercise 3

1) (a) $2e^{2x}\left(\sin 2x+\cos 2x\right)$ (b) $\dfrac{2x+5}{x^2+5x+1}$

(c) $4\sec(4x-\pi)\tan(4x-\pi)$ (d) $2x\sin\left(4-x^2\right)$

(e) $\dfrac{\ln x-1}{\left(\ln x\right)^2}$ (f) $\dfrac{1}{x\ln 4}$

(g) $4^x\ln 4$ (h) $2\tan 2x$ (i) $2\text{cosec}^2(1-x)\cot(1-x)$

2) (a) $\left(2\sec^2 2x\right)e^{\tan 2x}$ (b) $12\sin^2 4x\cos 4x$

(c) $-6\ \text{cosec}^2 3x\cot 3x$ (d) $f(x)=\dfrac{2\ \text{cosec}^2 4x}{\sqrt{\left(\cot 4x\right)^3}}$

4) (a) $\dfrac{-2x-y}{x+2y}$ (b) $0\ ,\ -1$ (c) $-\dfrac{2}{3}$

5) (a) $\dfrac{2}{\sqrt{2-x^2}}$ (b) $\dfrac{5}{\sqrt{1-25x^2}}$ (c) $\dfrac{e^x}{1+e^{2x}}$

7) $9\sec 3x\left(\sec^2 3x+\tan^2 3x\right)$

8) $\dfrac{3\ln x}{x^2}(2-\ln x)$

9) $\dfrac{-3x^2y^2-1}{1+2x^3y}$

10) $\dfrac{y\cos(xy)}{2y-x\cos(xy)}$

11) (a) $-2\sin x\cos x$ (b) $\dfrac{\pi}{2}$, $\dfrac{3\pi}{4}$

12) -2

13) $\left(\dfrac{1}{2}, \dfrac{1}{2e}\right)$

14) 3 cm/sec

15) (a) 76.9 m/s (b) 0.142 rad/sec

Exercise 4

1) (a) $\dfrac{3}{4}$ (b) $\dfrac{1}{2}$ (c) $\dfrac{1}{8}$

2) $\dfrac{1}{0}$ is the wrong format for L'Hopital's Rule

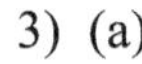

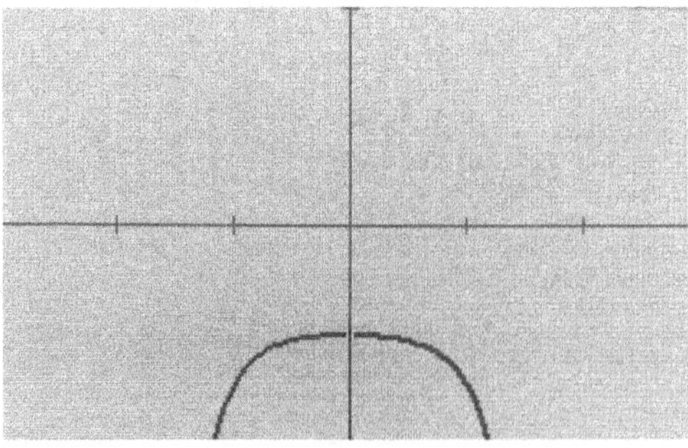

(c) (i) $-$ 0.500844 (ii) $-$ 0.500844 (e) $-$ 0.5

4) (a) $\frac{1}{0}$ is the wrong format to continue L'Hopital's Rule

(b) $+\infty$ (c) $-\frac{1}{e}$

5) (a) 0 (b) $\frac{1}{3}$

Exercise 5

1) (a) $2x^3+4x^2+3x+c$ (b) $\frac{-9}{x}+c$ (c) $\frac{-12}{\sqrt{x}}+c$

(d) $9\sqrt[3]{x^4}+c$ (e) $\frac{-3}{16x^4}+c$ (f) $3x^6+\ln x+c$

2) (a) $\frac{1}{2}\ln\left(x^2+1\right)+c$ (b) $\frac{1}{4}\sin 4x+c$ (c) $\frac{1}{3}\tan 3x+c$

(d) $\frac{1}{5}e^{5x}+c$ (e) $\frac{1}{4}\sin^4 x+c$ (f) $\frac{1}{10}e^{5x^2}+c$

(g) $\frac{1}{18}(3x+2)^6+c$ (h) $\frac{1}{3}\sqrt{(2x-1)^3}+c$ (i) $\sqrt{2x-1}+c$

3) (a) $\ln\left(3x^2+4x+5\right)+c$ (b) $\ln\sin x$

4) (a) $\ln 3$ (b) 1 (c) $\frac{3-2\sqrt{3}}{3}$ (d) $\frac{1}{2}\left(e^2-1\right)$

(e) 4 (f) 2

5) (a) 0 (b) 32

6) 0

7) (a) 320 m/s (b) 128 m

8) $\frac{1}{2}\ln\left(6+e^4\right)$

9) $\ln\sec x$

10) (b) 32 m (c) 64 m (d) 0 m

(e) $\int_0^4 12t-3t^2\ dt+\left|\int_4^6 12t-3t^2\ dt\right|$ (f) $\int_0^6 12t-3t^2\ dt$

11) $y=4x^2+3x+2$

12) $y=\ln\left(x^2+1\right)+2$

13) (a) 7

14) (a) $\ln x$ (b) $x\ln x-x+c$

15) (a) –30 (b) –28 (c) 2

16) (a) $\frac{1}{2}(1-\cos 2x)$ (b) $\frac{\pi}{12}-\frac{\sqrt{3}}{8}$

17) (b) $-\cos x+\frac{2}{3}\cos^3 x-\frac{1}{5}\cos^5 x$

18) (a) $\tan x-x+c$ (b) $\sin x-\frac{1}{3}\sin^3 x+c$

19) (a) $\frac{1}{2}(\ln x)^2$ (b) $\frac{1}{2}(\ln 5)^2$

20) $f(x)=3\ln x+x+4$

Exercise 6

1) (a) $x+4\ln(x-2)+c$ (b) $\frac{1}{21}(3x+5)(2x+1)^6+c$

(c) $\frac{2}{27}(3x+2)\sqrt{3x-1}+c$

2) (a) $\frac{1}{4}\sin 2x-\frac{1}{2}x\cos 2x+c$ (b) $\frac{1}{9}e^{3x}(3x-1)+c$

(c) $\frac{1}{4}x^2(2\ln x-1)+c$ (d) $x^2\sin x+2x\cos x-2\sin x+c$

(e) $\frac{1}{32}e^{4x}\left(8x^2-4x+1\right)+c$

3) (a) $\frac{2}{x-2}-\frac{2}{x+5}$ (b) $2\ln\left(\frac{x-2}{x+5}\right)+c$

4) $\arcsin\left(\frac{x}{4}\right)+c$

5) $\frac{1}{6}\arctan\left(\frac{3x}{2}\right)+c$

6) $\frac{1}{3}\arcsin(3x)+c$

7) $\frac{1}{35}\arctan\left(\frac{7x}{5}\right)+c$

8) (a) $\dfrac{1}{1+x^2}$

(b) $x\arctan x-\dfrac{1}{2}\ln\left(1+x^2\right)+c$

9) $x\arccos 3x-\dfrac{1}{3}\sqrt{1-9x^2}+c$

10) (a) $\dfrac{\pi\sqrt{3}-3}{6}$ (b) $\dfrac{2+3\ln 3}{4}$

11) (a) $\dfrac{1}{2}e^x\left(\sin x-\cos x\right)+c$ (b) $\dfrac{1}{2}\left(e^{\pi}+1\right)$

12) (a) $-(x+1)^2+4$ (b) $\arcsin\left(\dfrac{x+1}{2}\right)+c$

13) (a) $4\left(x-\dfrac{1}{2}\right)^2+16$ (b) $\dfrac{1}{8}\arctan\left(\dfrac{2x-1}{8}\right)+c$

14) (a) (i) $2x-4$ (ii) $\dfrac{1}{2}\ln\left(x^2-4x+3\right)+c$

(b) (i) $\dfrac{1}{2(x-1)}+\dfrac{1}{2(x-3)}$ (ii) $\dfrac{1}{2}\ln(x-1)+\dfrac{1}{2}\ln(x-3)+c$

Exercise 7

1) (a) $\ln\sqrt{2}$ (b) $\pi - \frac{\pi^2}{4}$

2) 9

3) (a) $e^2 - 1$ (b) $\frac{\pi}{2}\left(e^4 - 1\right)$

4) (a) $2\ln 2 - 1$ (b) $\pi\left[2(\ln 2)^2 - 4\ln 2 + 2\right]$

5) $\frac{512}{3}$

6) (a) $(-3.0465\ ,\ 0.0475)$ and $(0\ ,\ 1)$

(b) $\int_{-3.0465}^{0} \cos(0.5x) - e^x \, dx$ (c) 1.045

7) (a) $\pi\int_{1}^{2} (\ln x)^2 \, dx$ (b) 0.592

8) (a) $\frac{\pi}{7}$ (b) $\frac{3\pi}{5}$

9) $\frac{3\sqrt{3}}{2}$

10) (b) $\frac{128}{5}$ (c) 4096π (d) $\frac{256\pi}{3}$

Exercise 8

1) (a) $y^2 = 4\sin 2x + c$ (b) $y^2 = 4\sin 2x - 1$

2) $y = A\sqrt{x^2+1}$

3) (a) $y = \arcsin\left(\dfrac{x+c}{x}\right)$ (b) $y = \arcsin\left(\dfrac{x-1}{x}\right)$

4) $y = \ln\left(\dfrac{3x+c}{x^2}\right)$

5) (a) $\sin x$ (b) $y = \dfrac{1}{2}\sin x + k\csc x$ (c) $y = \dfrac{1}{2}\sin x + \dfrac{3}{8}\csc x$

6) $y = x^2 + \dfrac{c}{x}$

7) (a) $y = 2 + \dfrac{k}{e^{x^3}}$ (b) $y = 2 + ke^{-x^3}$

8) (a) $x = 50e^{kt}$ (b) 792 (c) 43.0 years

9) (a) 7.536 (b) $y = \dfrac{e^{2x} + 4e^3 - e^6}{2e^x}$ (c) 7.598

10) 1.573

Exercise 9

1) (a) $1 - \frac{x^2}{2!} + \frac{x^4}{4!} - \frac{x^6}{6!}$

(b) 0.6215993875 ...... (c) 4

2) $x - \frac{x^3}{3} + \frac{x^5}{5}$

3) $\ln 0$ does not exist

4) $1 - x^2 + x^4 - x^6 + \ldots\ldots$

5) $x^2 - \frac{x^6}{3!} + \frac{x^{10}}{5!} - \frac{x^{14}}{7!} + \ldots\ldots$

6) $1 + x - \frac{x^3}{6} + \frac{x^4}{24}$

7) (b) $-x - \frac{x^2}{2!} - \frac{x^3}{3!}$

8) (a) $y = 1 + x^2 + \frac{x^4}{2}$ (b) $y = 1 + x^2 + \frac{x^4}{2}$

9) $y = 1 + \frac{x^2}{2} - \frac{x^3}{6}$

10) (a) $y = 1 - x + \frac{3x^2}{2} - \frac{x^3}{2}$ (b) $y = 2x - 2 + \frac{3}{e^x}$

www.ingramcontent.com/pod-product-compliance
Lightning Source LLC
LaVergne TN
LVHW010103170826
845678LV00012B/2223
9798841079163